AF773104

NEW DIRECTIONS IN THE PHILOSOPHY OF TECHNOLOGY

PHILOSOPHY AND TECHNOLOGY
VOLUME 11

The titles published in this series are listed at the end of this volume.

OFFICIAL PUBLICATION OF
THE SOCIETY FOR PHILOSOPHY AND TECHNOLOGY

PHILOSOPHY AND TECHNOLOGY
VOLUME 11

NEW DIRECTIONS IN THE PHILOSOPHY OF TECHNOLOGY

Edited by

JOSEPH C. PITT
Department of Philosophy,
Virginia Polytechnic Institute
and State University, Blacksburg

KLUWER ACADEMIC PUBLISHERS
DORDRECHT / BOSTON / LONDON

A C.I.P. Catalogue record for this book is available from the Library of Congress.

ISBN 978-90-481-4603-1

Published by Kluwer Academic Publishers,
P.O. Box 17, 3300 AA Dordrecht, The Netherlands.

Kluwer Academic Publishers incorporates
the publishing programmes of
D. Reidel, Martinus Nijhoff, Dr W. Junk and MTP Press.

Sold and distributed in the U.S.A. and Canada
by Kluwer Academic Publishers,
101 Philip Drive, Norwell, MA 02061, U.S.A.

In all other countries, sold and distributed
by Kluwer Academic Publishers Group,
P.O. Box 322, 3300 AH Dordrecht, The Netherlands.

Printed on acid-free paper

TABLE OF CONTENTS

INTRODUCTION

The papers in this final volume of the *Philosophy and Technology* series in important respects are both an extension of earlier work in the philosophy of technology and excursions into new areas which hold promise for fruitful philosophical discussion.[1] In this sense, they represent philosophical work at its best while also serving as a fitting close to a series which over its complete history in various guises, always under the able editorship of Paul Durbin, has been the single most important vehicle for creating the field we know as the Philosophy of Technology. Philosophy is a dialogue and its participants come from both the present and the past. That philosophers are able to build on the work of their predecessors in fruitful and productive ways shows that the philosophy of technology is alive and well and ready to enter the mainstream of philosophical thought. What I mean by this is that philosophers of technology are increasingly interested in more than arcane exercises in Heideggerian logic or in mere social criticism, work which has value, but which speaks to a limited and self-selecting audience that has helped create the illusion that the philosophy of technology is a marginal area of interest.

The world of technology is the world we have created. In the past, much work in the philosophy of technology has been concerned with the impact (generally assumed to be negative) that "Technology" (writ large) has had on human values (usually invoked against the background of some ideal utopian state of human existence). Philosophers are now attending to how specific technologies affect our lives both physically and conceptually. This is often done without employing an ideological agenda, which makes the arguments advanced both more appealing and more accessible. It also shows that philosophers of technology have become more secure on their area of scholarship and no longer need to hide behind obscure language or Marxist rhetoric to justify their choice of research.

This is not to say that there has been a complete transformation in the philosophy of technology. Anyone familiar with the history of philosophy knows that philosophical problems are transformed over time. This is due to a number of reasons and, I would argue, generally cannot be predicted. Sometimes the forces of history demand that we ask and answer a different type of question.

Joseph C. Pitt (ed.), New Directions in the Philosophy of Technology, vii-ix.

This happened in the late 1960s and early 1970s when students caught up in the anti-war movement demanded that their teachers do more than spin convoluted and irrelevant ethical theories about a world which never existed and come to grips with the morality of a real war which by all accounts was unjustifiable. Sometimes philosophical problems are overtaken by scientific developments, as may be the case with the mind/body problem in the light of current work in cognitive science.

It also holds that fields do not completely change all at once. When I suggested that the philosophy of technology is moving into the mainstream as it leaves behind its more polemical and metaphysical past, this is not to say that all philosophers of technology have abandoned the past. Nor should they. But, likewise, those who cling to older ways of tackling the problems technology presents should not dismiss new questions, new techniques of analysis, new centers of attention. If the philosophy of technology is to be a dynamical feature of the philosophical landscape, it must try new approaches, reject old solutions, challenge entrenched assumptions, offer specific examples and arguments, in short, show that it is real philosophy. I propose that the papers in this volume represent the advance of the new wave and that as such they offer hope for serious integration of the philosophy of technology into the broader concerns of the philosophical world. Not everything new and exciting in the world of philosophy of technology is represented here. We are missing examples of philosophical dealings with engineering, for example. So, the papers here are not being broadcast as defining the new areas of research. They represent, by and large, examples both of good philosophical analysis and of new approaches to the philosophy of technology. Some avenues opened here may close quickly, while others may branch in unexpected and productive ways. Not every paper will become a classic. But if, taken as a whole, they advance our abilities as philosophers to ask and to answer in increasingly coherent ways questions about the world we have made, then they will have been successful.

A special word of thanks is required here. This volume would never have seen the light of day were it not for the devoted efforts of Karen Snider and the assistance of Lisa Huff. They have our deepest appreciation.

Joseph C. Pitt
Newport, Virginia

NOTE

[1] The papers presented here are a selection from among papers presented at the VIth International Conference of the Society for Philosophy and Technology held at Mayaguez, Puerto Rico, in March 1991.

JOSEPH C. PITT

DISCOVERY, TELESCOPES, AND PROGRESS

1. INTRODUCTION

In this paper I attempt a number of things. Together they constitute some first steps toward the development of a new research program. It is a proposal for a new way of conceptualizing the relation between science and technology. It is also an attempt to find a way to escape some old philosophical dichotomies; dichotomies which have kept philosophers of science and philosophers of technology apart. The underlying theme is this: following Derek Price, it seems clear that *progress in science is a direct function of increasing sophistication not merely in instrumentation, but in the technological infrastructure which underlies and makes mature science possible.*

Price claimed that, "historically, the arrow of causality is largely from the technology to the science" (Price, 1983), but it is only part of the story. By emphasizing the causal priority of technology in scientific progress, Price was attempting to overcome a popular characterization of the relation between science and technology in which technology is placed in a second class position, the offshoot of science or sometimes its "handmaiden." Price was on the right track, pointing out that despite the fact that historians and philosophers of science have a tendency to talk about progress in science in terms of the history of ideas, a significant role is played by technology, a role largely ignored by these same philosophers and historians of science.

The typical history of ideas story of science proceeds by relating that, for example, Newton's mechanics replaced Aristotle's and then quantum mechanics replaced Newtonian mechanics. The story is usually told in Kuhnian fashion, without any mention of the means by which anomalies were discovered.[1] It is merely announced that following a certain experiment, it was decided that so-and-so's theory was false and was replaced by another. Thus, a typical bad history would tell you that Michelson and Morley's experiment was developed to test for aether drift, as predicted by Newton's theory.

Joseph C. Pitt (ed.), New Directions in the Philosophy of Technology, 1-16.

Once it was discovered that drift did not occur, Newton had to be abandoned. Enter Einstein, and all is saved. Very few histories reveal that Newton did not talk about aether drift; the notion evolved over a hundred years in the course of his successors' efforts to adjust his theory in light of their experience with it. Likewise, very few accounts tell you about the details of the Michelson-Morley experiment.[2] The point here is that on the history of ideas account the history of progress in science is made to read like merely the replacement of one bad theory by another once the bad theory is discovered to be faulty. What is ignored is the role in all of this of *the technological infrastructure* within which the falsification and/or confirmation of theories takes place, to the extent that theories are falsified and/or confirmed.

More to the point, few scholars talk about the epistemology of experimentation, or the nature of the link between experiments and the theories they are supposed to test,[3] or the impact of experiment design and the availability of materials, techniques, and instruments. This aspect of the story of the progress of science/technology is important particularly at times of dramatic changes, such as are marked by the replacements of one major theory by another, because it is precisely at this juncture that what counts as evidence and how it comes to count as evidence is often at issue.[4]

In this paper, I turn explicitly to the role of the technological infrastructure of science in the growth of knowledge in general. I start by exploring some features of the manner in which Galileo's development and use of the telescope helped create an initial technological infrastructure for astronomy and then move to a sketchy reconsideration of that notion as it occurs in modern guise. In so doing I hope to make plain what is meant by a technological infrastructure of science. Instead of attempting to argue one side or another of the old science/technology debate, some of the issues are recast so as to demonstrate the epistemological importance of a technological infrastructure construed as interrelated sets of artifacts and structures. Furthermore, I want to suggest, if not argue here, that just as it makes no sense to talk broadly of technology, it makes no sense to speak of the history and development or importance of a single artifact, suggesting, as this does, that once invented artifacts remain stable over time. The thesis is direct: *the development of new information in a mature science is, by and large, a function of its technological infrastructure.* In short, scientific discovery today almost completely depends on the technological context without which modern science would be impossible. I will not raise the question of the merits of this situation until the end of my discussion, although I will provide a hint: in this age of increasingly theoretical science, the technology behind the science may be our only contact with reality, and even so it is at best a tenuous one. But now let us turn to the question of technology and discovery.

2. DISCOVERY

There has been little discussion of discovery by philosophers. To the extent that the issue has been raised it comes in three contexts. First, there are the problems the concept of discovery creates or sets for cognitive science. I will not discuss that at all. Second, discovery is a problem for realism – where the debate hovers over the distinction between discovery and invention. Consider the question, for example, whether it is possible to discover something which does not materially exist, that is, an idea or a theory. The problem of figuring out what this means rapidly becomes tangled, despite the fact that things seem fairly easy at first. One cannot invent, for example, the Americas – they are already there, so we discover what is there. But scientific theories are invented, not discovered, and yet they are supposed to be about what is there. To say we discover a theory makes it sound like the theory has been lying around waiting for us–but that is too Platonic for my tastes, especially since we keep "discovering" the wrong theories–i.e., false ones. On the other hand, we need to avoid making it sound like we invent theories out of thin air–surely scientific theorizing has some relation to what is there. Thus, there is a certain tension surrounding discussions of discovery in accounts of the development of scientific theories, which tension is generally resolved by invoking a temporal ploy–we *begin* by inventing ways of speaking about situations which have avoided our efforts to understand them until we have some sort of acceptable proof to the effect that what we invented to explain the situation really is there. At that point we say we have discovered these new phenomena, like gravity or quarks. But this is not really a solution; it is more like a wiggle.

The third situation, in which discovery has been a topic for philosophers of science, is as unhappy as the second. It is to be found in the context of Reichenbach's distinction between the context of discovery and the context of justification – a distinction employed so well by Popper in the *The Logic of Scientific Discovery*. Popper made things very difficult with his classic dismissal of discovery as an issue for philosophers by characterizing it as a fit topic only for psychology. His own view is frustratingly obscured through the mistranslation of the original German title, *Logik der Forschung* as *The Logic of Scientific Discovery*, when Popper rejects the very concept of a logic of discovery in the first 5 pages. Surely we would all have been served better if the title of Popper's book had been more accurately translated as *The Logic of Scientific Research* – for it was the structure of *that* process, "Forschung" in German meaning "research" or "investigation," with which Popper was really concerned. But, the follies of mistranslation to one side, it is nevertheless true that, for the most part philosophers of science in the middle years of this century bought into the Reichenbach/Popper view that discovery is not susceptible to logical analysis and, hence, is not an appropriate topic for discussion. It was only some years later, following the publication of Kuhn's *Structure of Scientific Revolutions*, when the locus of philosophical attention shifted to the historical process of science and away from concerns over its rational reconstruction, that discovery once again became an acceptable topic. Only now it posed problems of the second sort noted above–i.e., how does a scientific realist deal with the discovery/invention of theoretical entities?

This issue is currently, to a certain extent, a hot topic. It has taken on a slightly different shape, which is not unexpected given that many old problems never really die; they often reappear in a different vocabulary and context, wearing new clothes as it were. Today the invention/discovery battle is taking place between philosophers of science who are Scientific Realists and sociologists of science belonging to what is euphemistically known as the Strong Programme. Scientific Realists believe some version or other of the claim that the theoretical entities mentioned by our best scientific theories actually do exist. Thus, for Scientific Realists we eventually do discover the real world. There are varieties of realism but they do not concern us now. The practitioners of the Strong Programme, on the other hand, could be said to be inventionists, although they prefer the term "social constructivist." On their view, what most of us call the real world, indicating by that that feature of reality which is independent of us, is nothing more than the result of negotiation among scientists with special axes to grind. Reality is invented or constructed. Now I am not a Scientific Realist of the average sort, but this does not leave social inventionism as our only other alternative. I advocate a new form of realism, Sicilian Realism–a view I will return to below. While I am a firm believer in the social aspects of the creation of scientific knowledge, I cannot accept the view that reality is mere invention.

3. DEFINITIONS

If we are going to avoid old conundrums of the sort rehearsed above, and if we are to continue on to make some sense of the three notions I would like to address, discovery, technological infrastructure, and scientific progress, we should back off the old tracks and start somewhere fresh. First, let us consider some basic notions in the form of working definitions:

DISCOVERY:
the cognitive apprehension of that which has not been so apprehended or apprehended in that manner before.

TECHNOLOGY:
humanity at work.

THE TECHNOLOGY OF DISCOVERY:
humanity at work cognitively apprehending that which has not been so apprehended or apprehended in that manner before.

These definitions present a few problems deriving from the realist/constructivist debate. For example, do we, in cognitively apprehending electrons using an electron microscope for the first time, invent or discover electrons? The way to avoid getting stuck back in the very situation we are trying to avoid is to take our definitions seriously. The definition offered above makes no ontological claims, only an epistemological one. One must "cognitively apprehend something new or in a new way." It doesn't follow that such an act entails that what is cognitively apprehended must exist. Thus this account of discovery hopefully avoids the old problems of the realist and the constructivist, at least in the manner in which the groups mentioned above were plagued by them.[5]

Turning back to the definitions, I wanted to lay them out so as to help clarify some of the issues that are before us. But to understand the role of the technological infrastructure of science, we seem to have both too much and too little in these definitions. Attending to "cognitively apprehending people at work in a new way" is not going to help us explore the sense in which sets of artifacts generate new scientific discoveries.[6] We need something else, we need to know the manner in which further scientific work depends on new developments in the artifacts, i.e., an account of the *invention and modification* of the relevant artifacts in these circumstances. That is considerably more complicated. Second, we need to define "technological infrastructure."

> *A TECHNOLOGICAL INFRASTRUCTURE:*
> a set of mutually supporting artifacts and structures which enable human activity and provide the means for its development.

The notion of *mutually supporting* sets of artifacts is difficult to nail down in the abstract. What ultimately is perhaps most important is not the notion that science works within a framework of interrelated sets of artifacts, but the realization, nay, *discovery* that the technological infrastructure has itself grown and developed over time in conjunction with those features of the activity we call science. Thus I am not claiming that science, whenever and however it is or was practiced, has this kind of technological infrastructure. However, the development of a technological infrastructure is essential if science is going to continue to provide us with new discoveries about how the universe works. In short, after slow and modest beginnings, a *developed* science *requires* this kind of technological framework. The sorts of investigations and explanations it is called upon to produce require more than mere unaided human thinking alone can produce. I will return to consider the consequences of this claim later. For now, this is enough speculation; let's start to build the case.

4. GALILEO AND THE TELESCOPE

Modern science begins with the scientific revolution of the 17th century in which Galileo played a major role. He was an advocate for two of the major technological innovations which made the revolution possible, for the revolution involved more than rejecting the geocentric view of the world in favor of the helio centric. Crucial to the scientific revolution was the development of mathematical physics, or to put it in another way, the introduction of mathematics as an essential tool of science. Galileo was a prime contributor in building the case for the use of mathematics in physics (Pitt 1991). He was also, if not the inventor, at least the most successful early user of the telescope, the first to use it to explore regions of the universe previously inaccessible to us, to use the results of those explorations to challenge established theories, and as support for new ways of thinking.

I have told the story of the way in which Galileo was captured by the telescope before (Pitt, 1987), but let me rehearse it again briefly. There is a temptation to view Galileo's future relationship with the telescope as almost inevitable, but I will resist it. Almost fifteen years prior to learning about that device he had already demonstrated his knack for working innovatively with instruments. In 1592 Galileo was appointed to the chair of mathematics at Padua. Shortly thereafter he began giving courses in private on material outside the normal curriculum in order to supplement his income. Somewhere around 1595, following such a course on military architecture, he developed the first version of what came to be known as his military compass, a device for measuring distances and altitudes and for which he composed his first published work, appearing in 1597. By 1599 he had perfected the device into an all purpose calculator, which according to Stillman Drake "was capable of solving any practical mathematical problem that was likely to arise–swiftly, simply, without requiring previous mathematical education, and sufficiently accurate for ordinary practical purposes" (Drake, 1978, p. 9). But, in what is beginning to emerge as a familiar pattern for Galileo, the basic idea was "borrowed" from a friend of his, the Marquis Guidobaldo del Monte. I say this is a familiar pattern because around the same time, 1595, he appropriated the basic scheme for what was to become his ill-fated theory of the tides from yet another friend, Fra Paolo Sarpi, and in 1609, upon hearing that a Dutch lensmaker had made a device which could bring distant images near he sat down to reconstruct it, checking first with his good friend Sarpi to see if it was feasible and then hastening to beat out an itinerant peddler who was on his way to Venice with a Dutch version. Galileo had already realized that such an instrument would be of value to the Venetian navy to warn them against pirates and he also thought he could achieve some financial advantage by building one and giving it to the Doge of Venice, who was nominally his employer at the time. He succeeded in manufacturing an appropriate device and managed to get it into the hands of the Doge first. But his ploy was only partially successful. The Doge was impressed and ordered Galileo's salary as a professor of mathematics at Padua doubled to 1000 florins. But the small print in the contract said that Galileo would also not receive another raise for life.

Miffed at being finessed, Galileo had occasion to show his new instrument to Count Cosimo d'Medici when he was home visiting Florence later in the year. Together they tried it out on the moon and discussed the possibility that the dark spots were shadows caused by mountains. But Galileo's eight power telescope was not strong enough to resolve the issue. And now, for our purposes, the crucial events begin to unfold. When Galileo returned to Padua he built a 20 power telescope, confirmed his suspicions about the shadows on the moon, wrote the Count, negotiated a deal, and moved home to Florence to take up his new post as the resident mathematician and philosopher to the Medici. Now free of the restrictions of having to teach a curriculum dictated by the Church, or having to teach at all, and urged on by Cosimo's own interests, Galileo continued to make telescopic observations of the moon and then of Jupiter, discovering its moons. He also became Italy's premier manufacturer and supplier of telescopes, adding to his instrument business. Word of Galileo's discoveries spread, placing him under pressure to publish his findings before he was scooped–priority of discovery being as important then as now. When he finally published *The Starry Messenger* in 1610, the fate of modern astronomy was sealed.

Let me take a minute to defend this rather dramatic claim. Galileo was not the first to use an instrument to investigate the heavens. The astrolabe, a device for determining the positions of the planets and the stars, already had a long and rich history.

The quadrant was also a device used to determine positions in the heavens. But unlike the astrolabe or the quadrant, the telescope produced fundamentally new kinds of information. The telescope did not, as did the astrolabe, merely assist in the refinement of measurements according to an established theory. It produced fundamentally new information about the structure and population of the heavens. It forced a transformation in cosmology. The instrument, in effect, required a major overhaul of theory. What was being demanded of theory then forced a reworking and refinement of the instrument, which in turn pushed the matter even harder toward theory revision. A basic new feature had been added to the activity of science–the interplay between instruments and theory.

Later, but not too much later, the single instrument was to become a complex of instruments. Galileo originally intended his telescope to be handheld for maritime use. But for astronomical purposes it needed a base, then a fixed position from which the observations could be regularized. Tables of sightings could now be corrected, and the need for further refinements in the tables would force refinements in the telescope itself.

For example, a major problem in astronomy was determining the size of the planets. For this, Galileo's telescope with its concave len was not the optimal instrument. In the 1630s it began to be replaced by what van Helden (1989, p. 113) calls "the astronomical telescope" which had a convex ocular and produced greatly improved clarity in its images. It also had a broader field of vision which permitted the introduction of a micrometer into the instrument itself, thereby improving the precision of measurements. This was the kind of instrument Huygens used to measure the diameters of the planets. Slowly, Galileo's simple device was becoming a set of things, each part of which could be refined and in so doing would ramify its effects on the others; perhaps not all the others all the time, but a kind of domino effect was evident. Furthermore, the availability of increasingly precise measurements of particular features of the observable universe also forced changes in the manner in which the relative distances of the planets was calculated. So now we have the instruments and their refinements forcing changes not only in cosmology but in the auxiliary methods which augment it. In this manner the discovery of the size and structure of the solar system and then the universe was undertaken.

The story could be told without mentioning the instruments. For example, we could say,

Galileo showed there was more than one center around which planets revolved, forcing a revision of the geocentric theory of the universe. His methods were developed in such a way as to allow for the determination of the distances between planets and the relative sizes of the planets. Modern astronomy continues his program of empirical investigation of the universe.

That says what we have been saying, but the picture it provides of science is, to say the least, impoverished. The mechanism behind the changing ideas is lost, without which we truly have no explanation.

Perhaps the need for a refined explanation is the proper motivation for including the technological infrastructure of science in our history of cŭlture. And so, by returning to explanation and its associated difficulties, perhaps I have failed to completely extricate us from the old problems. But there is some progress evidenced here. For if we want an explanation for the development of science, we need to offer more than a recitation of the sequence of ideas produced by scientists. We need an account of how those ideas were developed and why they were abandoned and/or refined. We are thus dealing with an issue in historiography. An *explanation* of scientific progress and discovery requires appeal to some mechanism. That is why the history of ideas approach is inadequate. I am proposing that *the mechanism* which makes the discoveries of science possible and scientific change mandatory *is the technological infrastructure* within which science operates. In short you can no longer do philosophy of science, history of science or even sociology of science without the philosophy and history of technology.

5. THE NEW ASTRONOMY AND ITS TECHNOLOGICAL INFRASTRUCTURE

Perhaps one more look will help us make the case more convincing. Optical astronomy has come a long way since Galileo's little eight power handheld telescope. We don't need to turn to the Hubble space-based telescope to see that. Not only have telescopes grown in size, but the necessary support systems have become more complicated. The truly large telescopes require massive housings, highly sophisticated background technologies to produce the machines and lenses, electricity to run the equipment and, once cameras were introduced, all the apparatus needed for quality nighttime photography and the optical theories to support interpretations of the products, computers to calculate position, manage the photography and coordinate the systems.

Astronomy is the science of the heavens. Its function is to describe the universe in terms of the relative constitution of and positions of its parts. To accomplish this goal astronomers need to be able to see the heavens. And so we have the elaborate technological infrastructure of the optical telescope. But to assume that the components of the universe are limited to those which can be seen by the human eye is absurdly homocentric. So if you add to the optical infrastructure, radio telescopes and theories upon which they are based, spectral telescopes, the use of high speed computers to not only control the telescopes, but to generate and interpret at least the first and second order information they generate, the computers and the computer programs necessary for all that, the launching of space based telescopes and the technological systems behind that, the infrastructure behind the computers, etc., and the list goes on–if you add all that in, the technological infrastructure of astronomy appears to swamp the goal of the science. But there is more, for at each stage, the development of the instruments is constrained by the fit with other instruments and the theories with which they interact and sets of instruments and their backup systems. The results of employing these systems forces restructuring of theories all the way down the line. It isn't just that new observations force revisions in the description of the heavens. The questions include how do you integrate spectral telescopy with optical? Do the theories behind the instruments cohere? How do anomalous results from one instrument, e.g., excessive red shift, affect the other theories?

We look with awe at the picture which the new space probe, appropriately called Galileo, sent of the Earth on its way to Jupiter. If we think about the technological infrastructure behind the pictures, we get some sense of what is involved. The pictures are not simply sent from the space vehicle, traveling at high speed in its own trajectory, but they are sent to earth, which is also traveling at high speed and on its trajectory. The "pictures" are transmitted as electronic code. That means they have to be disassembled, sent, reassembled, etc. The machinery, the programming and the capacity for mistakes is enormous. If you add the testing of scientific theories to the problem, and the interaction between the theories and the technological infrastructure, as well as among themselves, there can never again be a simple history of the ideas of science, nor should there be.

If the science is astronomy, or even cosmology, and once we understand what it takes to do cosmology today, then we must turn to the technological infrastructure to understand its results. It is no longer possible to say "science tells us . . .", and it is certainly misleading to say "science and technology tell us . . ." for no one has taken the time to spell out what that means. When we do spell it out we will find that what we really wanted to say was, "The technological infrastructure within which scientific theories are being developed and transformed makes it possible for us to describe and explain the universe in the following way." This contextualization of our science is extremely important. The kinds of things we come to know about the universe. Or to put it more dramatically, the universe modern science reveals to us, is a function of this complex interaction between theory and technological infrastructure. Furthermore, it would seem that with a different technological infrastructure "science" would yield a different universe, or would it? As we attempt to answer this question we also find ourselves back in the realism debate. Let us then turn briefly to realism one more time.

6. SICILIAN REALISM AND TECHNOLOGICAL INFRASTRUCTURES

If we take as our starting point the fundamental claim of scientific realism–namely, that theoretical entities are real, however glossed–and couple it with the historical awareness that theories change and are replaced, we have a problem: which theoretical entities from which theories are really real? A Sicilian Realist will say that they all are. Sicilian Realism is realism with a vengeance; the universe is a very complicated place, to echo Marjorie Grene and Richard Burian. What we manage to do with one theory/technological infracture is to cut the universe at one of its many joints. Optical telescopes tell us planets and stars are real. Radiotelescopes tell us there is more out there. Sicilian Realism admits all of this. What Sicilian Realism does not admit as at all necessary is the kind of reduction which normal realism assumes. Thus, atoms, electrons and quarks are all equally real–without one having to be reduced and explained by another. Seeing the universe in terms of atoms is a function of cutting it only one way, and there are others.

What we have to face is the fact that while there is no one necessary way to investigate nature, the mechanisms–read "technological infrastructures"–we develop to assist us set a complicated process in motion in which imagination and creativity is sparked and fed by the interplay between idea and artifact. Artifacts stimulate us to seek uses for them: how to couple them with other artifacts; how to interpret the results. Given different sets of artifacts–by definition different stimulations–we get different results. But we start small and go large in quick order. Compare Galileo's simple telescope with the complex that we need for a modern mountain-top observatory.

What are the consequences of accepting this characterization of the role of technology? Is it not the case, as I am sure some determinist will be sure to suggest, that that means not only that society is run by technology, but now science is too! No, that is not the proper conclusion to draw. It is not a question of which disembodied and reified nonentity, science or technology, controls anything. What a careful look at history will show is that as instruments are made more complex by individuals with specific objectives in mind (objectives sometimes, but not always, generated by theories), a complex of interrelated activities develops through which by choosing certain ways to augment the technological infrastructure, certain options are opened or shut for theoretical testing and exploration. People still make the choices, and they may choose badly, taking us down a dead end. Or they may opt for a system which does not have the backup to support it. This is what happened to the 19th-century astronomer, William Herschel. He built a forty inch telescope which was certainly a technological marvel. Only there were severe problems. The mounting for it proved unstable. The mirror was made of metal and lost its reflective capacity. It fell into disuse.

My point is that if you want to explain the changing claims and face of science, you have to go beneath the ideas to the technological infrastructure and then you have to look at unraveling the interactions between its parts and the mass of theories with which it is involved. It is that complex that makes it possible to apprehend new things or to apprehend things previously known but in a new way. The discovery of the structures of nature is a function of this complicated mutually interacting set of artifacts, ideas, systems and, of course, men and women. Telling that story puts us in a position to finally understand the nature of the scientific enterprise and how it generates new information. It should also alert us to the sensitivity of the system. With so much depending on so much, there are many opportunities for things to go wrong. Likewise, because of the complex of interdependent relationships, the determination of the accuracy or even of the import of the new information this system generates is not an easy task. Scientific knowledge becomes more tenuous and more dearly bought as the technological infrastructure grows, but it is increasingly impossible without it. Sometimes all we know is that this or that system works; we may not know what it is telling us.

Virginia Polytechnic Institute and State University

NOTES

1 A classic example of this can be found in a recent (1987) issue of *Mosaic,* an official NSF publication:

Every so often, in the long course of scientific progress, a new set of ideas appears, illuminating and redefining what has gone before like a flare bursting over a darkened landscape. It happened when Galileo realized that physical laws needed to be written with numbers and invented the scientific method, when Darwin found an entirely different way to consider the evolution of living things, when Freud placed consciousness and emotion in a new context, when Einstein found a radical way to look at space and time, and when Wegener launched an earth science based on continental drift.

2 Or that both experimenters were Americans and that the experiment was carried out in Cleveland, Ohio, at what was then the Case Institute of Technology. After all, with names like "Michelson" and "Morley" they just had to be British and the experiment must have taken place at the Cavendish; didn't they all?

3 This situation is changing. See Franklin (1986), Ackermann (1985), Hacking (1983), Galison (1987) and Cartwright (1989).

4 I have discussed some of these issues elsewhere, most recently in my (1991).

5 Only if your definition of "knowledge" entails existence would you be back in the old ditch in a hurry. Definitions of knowledge which entail the existence of the things which are known usually invoke a truth condition–such as in "knowledge = justified true belief." Luckily, there exist accounts of knowledge which avoid the problems truth conditions present. For example, on my account, which I will not belabor here, I distinguish between what is proposed by individuals as candidates for knowledge and the endorsement of those claims by the appropriate social community. An individual may think he or she has found the truth about a particular matter, but thinking or wishing so doesn't make it so. Only when the claim has been endorsed by a particular community does it count as knowledge. The criteria the community invokes may have nothing to do with truth–it may, for example, remain satisfied with coherence or with practical efficiency. But, and this is what counts here, if the community determines knowledge, then inevitably truth will go by the board (Pitt 1983). This is the germ which the social constructivist and most relativists exploit.

6 But it may bear on the resolution of discipline specific problems by importing techniques and individuals from other disciplines.

REFERENCES

Ackermann, Robert (1985). *Data, Instruments, and Theory.* Princeton: Princeton University Press.

Cartwright, N. (1989). *Nature's Capacities and Their Measurement.* Oxford: Clarendon Press.

Drake, Stillman tr. (1978). *Galileo Galilei, Operations of the Geometric and Military Compass.* Washington, DC: Smithsonian Institution Press.

Fisher, Arthur (1991). "A New Synthesis Comes with Age," in *Mosaic*, vol. 22, number 1, p. 3.

Franklin, Allan (1986) *The Neglect of Experiment.* Cambridge: Cambridge University Press.

Galison, P. (1987). *How Experiments End.* Chicago: University of Chicago Press.

Hacking, I. (1983). *Representing and Intervening.* Cambridge: Cambridge University Press.

Kuhn, T. (1962). *The Structure of Scientific Revolutions.* Chicago: University of Chicago Press.

Pitt, J. (1987). "The Autonomy of Technology," in *Technology and Responsibility*, edited by Paul T. Durbin, pp. 99-114.

_____ (1983). "The Epistemological Engine," *Philosophia*, 32, pp. 77-95.

______ (1991). *Galileo, Human Knowledge and The Book of Nature; Method Replaces Metaphysics.* Dordrecht: Kluwer.

Popper, K. (1963). *The Logic of Scientific Discovery.* London: Hutchinson.

Van Helden, Albert (1989). "The Telescope and Cosmic Dimensions," in *Planetary Astronomy from the Renaissance to the Rise of Astrophysics*, Part A, edited by R. Taton & C. Wilson. Cambridge: Cambridge University Press.

PETER KROES

TECHNOLOGY AND SCIENCE-BASED HEURISTICS

1. INTRODUCTION

The view that equates technology with applied science, that is, that blueprints for technological inventions can be straightforwardly derived from scientific theories or scientific knowledge in general, is historically and philosophically problematic.[1] Historians of technology tend to reject this view because it makes science the principal dynamical factor behind the development of technology, and therefore leaves almost no room for social influences. Philosophers of technology have severely criticized this model, because it naively assumes that technological designs can be deduced from scientific theories by simply feeding those theories with appropriate boundary conditions. More generally, they claim that technological knowledge is different in nature from scientific knowledge.[2]

The idea that technology is applied science is an integral part of the so-called *linear model* which postulates a sequential path from scientific knowledge to technological invention and innovation. If we add to this scheme the assumption that science follows an autonomous development according to a kind of internal dynamics (or logic), then we end up with some form of *technological determinism*.[3]

If it is agreed that there is no compelling logic leading from scientific knowledge to technological designs and that technological and scientific knowledge have different characteristics, how does science then contribute to technology? In this paper, I shall restrict myself mainly to one aspect of this problem, namely the influence of scientific theories on technological development. I shall propose an approach that is focused on the heuristic value of scientific theories for the solution of technological problems. This approach interprets technological development primarily as constituted by heuristic search processes.[4] Whenever a technological problem emerges, a heuristic search process is started with the intention to solve this problem. The basic idea put forward here is that one of the

Joseph C. Pitt (ed.), New Directions in the Philosophy of Technology, 17-39.

important ways science contributes to technological development is that it provides very powerful heuristics for the solution of technological problems in the context of design, construction and optimal use of technological artifacts.[5]

The notion "heuristics" will be used here in the sense of the whole complex of ideas, rules and guidelines that *direct* the search for the solution of technological problems. Heuristic rules spring from an analysis of all available information about the problem under consideration. The main function of heuristics is to generate strategies for problem solving. Generally speaking, heuristic rules will restrict the search for a solution to a few promising alternatives. It may also contain specific guidelines as to how these can be realized. There is (almost) always an element of uncertainty involved when heuristic rules are followed, that is, it is not clear beforehand whether or not heuristic rules will be successful or effective. That depends on the kind of problem involved and the information available. For familiar problems, heuristic rules that were successful in the past may have precipitated routine procedures that more or less guarantee the finding of an effective solution. It is, however, difficult to assess the value of heuristic rules employed in situations where novel problems have to be solved about which information is fragmentary and incomplete. In those situations, the uncertainty will be great.

Contrary to the technology-is-applied-science model, the approach taken here does not degrade technology into some kind of a derivative of science. It starts from the idea that the solution of technological problems is often *guided* not *dictated* by scientific knowledge. In other words, one of the important links between science and technology is to be found at the level of heuristic considerations which guide the solution of technological problems and the development of technological research programs. The approach taken here leads to a much less rigid connection between science and technology and leaves ample room for an autonomous development of technology *vis a vis* science. Even if it would be assumed that science develops in an autonomous way according to some inner logic–by itself a highly problematic assumption–that assumption would not, on our account, imply some form of technological determinism.

Our approach makes it possible to trace in detail the influence of scientific knowledge on technology and to do justice to one of the main factors determining the development of modern technology. This will be illustrated in the following section with two examples. The first analyzes schematically the role of scientific knowledge in heuristic rules employed in the nitrogen fixation process developed by Fritz Haber (1868-1934) and Carl Bosch (1874-1940) at the beginning of this century.[6] The second examines the heuristic rules for steam technology contained in Carnot's theory of heat engines. In both examples, fundamental theoretical insights had a great impact on technological development. The Carnot example is of special interest, because in this case the fundamental theoretical insights were developed starting from a technological problem. In the discussion I shall indicate some general topics which deserve special attention from the point of view of this "heuristic model" of the interaction between science and technology.

2. EXAMPLES OF SCIENCE-BASED HEURISTICS IN TECHNOLOGICAL PROBLEM SOLVING

2.1 *The Haber-Bosch Nitrogen Fixation Process*

The nitrogen fixation process is an interesting example for studying the science-technology interaction, because the Heat Theorem discovered by Walter Nernst (1864-1941) in 1906, today better known as the third law of thermodynamics, played a vital role in its development.[7] In the following, I shall analyze this role in terms of heuristic rules which were derived from this law and guided the search for a technological process for the production of ammonia. I shall first have a brief look at the Heat Theorem itself.

The Heat Theorem, in its modern formulation, says that in the neighborhood of the absolute zero point of temperature, changes in the entropy S become infinitely small:[8]

$$\lim_{T \to 0} dS = 0$$

The theorem implies that it is impossible to remove all the heat from a system, that is, to cool it down to the zero point of absolute temperature.

The Heat Theorem was an important addition to the two fundamental laws of phenomenological thermodynamics.[9] Nernst considered it to be the final completion of this theory. Together, the first and second law did not allow in all cases a complete determination of all thermodynamic quantities on the basis of thermal constants only, such as specific heat, evaporation heat, reaction heat, etc. This problem was solved by the third law. In particular, the Heat Theorem made it possible to determine completely the equilibrium state of a chemical reaction on the basis of thermal constants. Nernst, in other words, was able to compute the equilibrium state of ammonia gas for any temperature and pressure.[10] This fact had important consequences for the development of a technical process for producing ammonia, to which we turn now.

During the 19th century many attempts were undertaken to produce ammonia directly from nitrogen and hydrogen, for instance by way of electric discharges in nitrogen-hydrogen mixtures.[11] Although traces of ammonia were obtained, these attempts were not successful from a technological point of view. The technological problem of producing ammonia in large quantities was finally solved by Haber and Bosch at the beginning of this century. In 1903 the Margulies brothers of the Österreichischen Chemischen Werken approached Haber asking him to advise them in their attempts to produce ammonia. Less than a decade later, large quantities of technically synthesized ammonia appeared on the market for the first time in history.[12] Science contributed in a significant way to this technological success.

At the time that Haber started his work on the synthesization of ammonia, it was already known that the chemical reaction between nitrogen and hydrogen was an equilibrium reaction:

$$3H_2 + N_2 \rightarrow 2NH_3$$

It was also known that the equilibrium state depends on pressure and temperature: increase of pressure shifts the equilibrium to the right (more ammonia), whereas increase of temperature shifts it to the left. Exact data on the equilibrium state were, however, still lacking. The first thing Haber undertook was a measurement of the equilibrium state at very high temperatures (about 1000°C). At these temperatures the percentage of ammonia in a nitrogen-hydrogen mixture is extremely low (about 0.01 vol.%); this makes it very difficult to obtain reliable data about the equilibrium state. Haber, however, was forced to experiment at those temperatures because at much lower temperatures (where the percentage of ammonia is considerably higher) the reaction velocity is extremely low. Starting from his measurements, Haber could compute, using the law of mass action and an equation derived by Van't Hoff, the equilibrium state at other temperatures. For a stoichiometric mixture he found:

Temperature (^{o}C)	27	327	637	927	1020
Vol % NH_3	98.5	8.72	0.21	0.024	0.012

Table 1

These results were not very encouraging for the technical production of ammonia. That would require at least a percentage of ammonia of about a few percent; that is, the production would have to take place at a temperature of about 300 to 400°C. At those temperatures, however, the reaction velocity was much too low. And at temperatures where the reaction velocity was high enough (about 1000°C) the percentage of ammonia was much too low. As so often in technology, Haber was facing here two *conflicting* heuristic rules: from the point of view of the reaction velocity, the temperature had to be increased, whereas from the point of view of the equilibrium state of the reaction, the temperature had to be as low as possible. He tried to solve this problem by the use of catalysts. The search for good catalysts was to become of enormous importance later on in Haber's research program.

Around 1907 a controversy arose between Haber and Nernst about the reliability of Haber's measurements. According to Nernst, they could not be correct because they were not in agreement with calculations of the equilibrium state of ammonia gas based on his Heat Theorem. Haber's measurements were too high by a factor of three. In order to get reliable quantitative results, Nernst decided to investigate the equilibrium state of ammonia gas at high pressures (40 to 70 atmospheres); under those conditions, the percentage of ammonia is considerably higher and so it is possible to determine the equilibrium state more accurately. His experimental results showed a better agreement with the theoretically calculated values. Haber, challenged by Nernst, then also started to study the equilibrium state at high pressures; for some time Nernst and Haber quarreled about the interpretation of the measurements, but finally it appeared that Haber's results also strongly supported the validity of the Heat Theorem. By 1908 the situation was such that reliable data about the equilibrium state of ammonia gas and a theoretical underpinning of those data were at hand. Nevertheless, technical production of ammonia was still far away.

Nernst's controversy with Haber had peculiar consequences for the technical production of ammonia. On the one hand, prospects became worse, because Haber's original figures with regard to the percentage of ammonia had been too high. On the other hand, in his attempt to verify the Heat Theorem Nernst managed to realize an ammonia yield of almost 1% at a pressure of 50 atmospheres (with manganese as catalyst); this percentage was about a hundred times higher than Haber had achieved.

From these developments Haber drew the conclusion that, if the technical production of ammonia was possible at all, it would have to take place at high pressure. Before, he had regarded this option as unrealistic because he thought that in practice the construction of reactor vessels that would be able to endure such high pressures and temperatures would be impossible. Nernst had shown that, on a laboratory scale at least, such reactor vessels could be constructed. The construction of full-size high-pressure reactor vessels that could operate at high temperatures was, later on, a crucial step towards the industrial production of ammonia. It was mainly taken by Carl Bosch who played an instrumental role in the solution of many fundamental construction problems.

After 1908, Haber focused his attention strongly on the problem of how the reaction velocity could be increased with the help of catalysts. An effective catalyst would allow him to perform the synthesization of ammonia at lower temperatures where the production of ammonia was higher. In the beginning, osmium powder, a metallic element of the platinum group with a high melting point, was used; later on, uranium was used. With these catalysts Haber finally succeeded in developing a process at laboratory scale that was to become the basis for the grand-scale technical production of ammonia.

Beside the work of Nernst and Haber on the equilibrium state of ammonia gas, the discovery of good catalysts and the construction of suitable reactor vessels were key events on the road to technically realizable ammonia production. Of course, many other problems had to be solved, such as problems with regard to scaling up the process (a factor of 1000 from the experimental set up to the production unit), the poisoning of catalysts, the production of hydrogen, the development of Haber's "recycling" principle, etc.[13] We will not go any further into these aspects but instead will analyze the development of the Haber-Bosch process from the perspective of the science-technology issue.

How did science contribute to technology in this particular case? Clearly, it would be strongly misleading to consider the Haber-Bosch process to be simply an application of a theory of physical chemistry, namely the thermodynamics of equilibrium reactions. It would be an overstatement of the role of science. Thermodynamics did contribute to the clarification of the equilibrium state of ammonia gas under various conditions, but knowledge of the equilibrium states was clearly not sufficient to reach the technological aim. By 1908 most experimental and theoretical problems about the equilibrium state had been solved, but from a technological point of view no real progress had been made:[14]

> Theoretically, things were thus largely clarified; yet, still not a single step forward had been made with regard to *application in technology.*

It would be no less misleading to play down the role of science. It cannot be denied that the research conducted by Nernst and Haber was technologically relevant. According to Mittasch[15] it laid the foundations for the technical production of ammonia. In his opinion, the technical production of ammonia would not have been possible without the rapid development of physical chemistry at the end of the 19th century.[16]

The technical relevance of the investigations by Nernst and Haber of the equilibrium states of ammonia gas lies primarily in the fact that these investigations offered an insight into the physical/chemical possibilities of technical ammonia production. The results of this research were exact, quantitative data about the equilibrium states of ammonia gas as a function of pressure and temperature. On the one hand, these showed that the technical production of ammonia would not be feasible at atmospheric pressure and a temperature of 1000°C because of the experimentally determined low percentage of ammonia; the Heat Theorem, moreover, implied that at those conditions the percentage of ammonia could not be increased so that it would become technically interesting. Theoretical calculations based on the same Heat Theorem, on the other hand, showed that at a pressure of 50 atmospheres and a temperature of 500°C the percentage of ammonia would in principle be sufficient. Here, the "in principle" clause hides a major problem, for under those conditions the velocity of the reaction is from a technological point of view too low. This obstacle somehow had to be overcome to bring technical ammonia production within reach. In order to remove this obstacle, Haber studied extensively the effect of all kinds of catalysts on reaction velocity.

Thus, the application of the thermodynamic theory of equilibrium reactions to ammonia gas clearly pointed out in which direction a possible solution for the technical production of ammonia *might* be found. That was in this case the primary function of scientific knowledge. It directed attention to the search for an appropriate catalyst. In this case, the technical relevance of science showed itself not by way of a direct and concrete application but by way of heuristics.

A brief look at Haber's search for a good catalyst shows the value of a powerful heuristic. At the beginning of this century little was known about the manner in which catalysts affect the velocity of chemical reactions. This means that Haber and his assistants, in the absence of a strong heuristic, were more or less groping in the dark in their search for an effective catalyst. In such a situation, blind trial and error is the only means to proceed. Indeed, the figures are telling: until the beginning of 1912, 6500 experiments involving more than 2500 catalysts were performed. In 1919 these figures were respectively 10000 and 4000.[17] Of course, they formulated hypotheses about groups of elements and compounds that might have good catalytic properties, but these hypotheses and the heuristic rules based on them were not very reliable. In contrast, a few experiments on equilibrium states of ammonia gas combined with theory were sufficient to draw significant heuristic conclusions about the conditions under which technical ammonia production might be possible. In view of the lack of knowledge about the operation of catalysts, it is no surprise that the discovery of a good catalyst happened more or less by chance. In March 1909 Haber found out that osmium was a good catalyst. But according to Mittasch:[18]

It was a happy coincidence that precisely the form in which that metal was offered to him has an exceptionally good catalytic effect, that is, has a much better effect than osmium that is made in another way. Specifically, Haber received from the Auergesellschaft, with which he was in contact, an osmium powder that for their own purposes (osmium lamps) was produced in a very fine "amorphous" condition from amino compounds according to a special, secret process. As Haber very soon observed, this form is considerably more effective than the metal which is obtained, for instance, by reduction at higher temperatures using hydrogen, which is not much more effective than manganese.

A little bit of luck and some "Fingerspitzengefühl" had to compensate for the lack of a powerful heuristic in the search for a good catalyst!

2.2 *Carnot's Theory and the Efficiency of Steam Engines*

At the beginning of the 19th century, steam technology was in a state of intense development. New types of steam engines, based on the use of high pressure steam, such as the famous two-cylinder compound Woolf engine working at steam pressures of about 3 atmospheres, appeared on the market.[19] The manufacturers of these high pressure engines claimed that they were more efficient than, for instance, the traditional low pressure engines, such as the Watt engine. It soon became a matter of dispute whether or not these engines were indeed more efficient and whether or not the efficiency of steam engines could in principle be improved by the use of high pressure steam.[20] It was also often suggested that the performance of steam engines could be considerably increased by replacing steam by other vapors or gases.[21] From a theoretical point of view, little was known at that time about the relation between the efficiency of steam engines and the pressure of steam[22] or about the effect on efficiency of choosing another medium for the transport of heat from the boiler to the condenser. In this respect, Carnot's analysis of heat engines in his *Reflexions on the Motive Power of Fire and on Engines Fitted to Develop That Power* (1824)[23] was a major step forward in the theoretical understanding of the efficiency of steam engines. In it he laid the foundations for the science of thermodynamics and for the modern thermodynamic conception of the efficiency of steam engines.

In the following, we shall first have a look at the modern notion of efficiency and compare it with the notion of efficiency developed by 18th century engineers. Then we will briefly turn to one of the main heuristic principles guiding the attempts to improve steam engines during the 18th century. After a brief digression into the nature of Carnot's work, its heuristic impact on steam technology will be discussed.

From a modern, that is, thermodynamic point of view, a steam engine is an energy converter: it transforms energy in the form of heat Q into energy in the form of mechanical work W by transporting heat from a heat reservoir with temperature T_2 to a heat reservoir with temperature T_1 ($T_2 > T_1$). For a steam engine, these reservoirs correspond respectively to the boiler and condenser. Not all the heat Q is transformed into mechanical work W; the ratio W/Q is called the thermal efficiency of a steam engine. According to thermodynamics, there is a maximum efficiency with which the conversion of heat into work can take place. This maximum is reached in an ideal Carnot process, and is given by:

$$\frac{W}{Q} = \frac{T_2 - T_1}{T_2}$$

with Q the amount of heat taken from the reservoir with temperature T_2. No heat-to-work transformer working between the same heat reservoirs can have a greater efficiency. Sadi Carnot was the first to develop the idea that there exists a maximum to the thermal efficiency in his *Reflexions*.

By the time Carnot published his booklet, steam technology was more than a century old. From the very beginning of the use of steam engines for the drainage of mines, their efficiency had played an important role. The early steam engines were notorious for their consumption of fuel (mainly coal). Especially in areas with a short supply of fuel, steam engine engineers were very much efficiency minded and tried to minimize the consumption of fuel. But the notion of efficiency employed by engineers during the 18th century and the beginning of the 19th was completely different from the above thermodynamic notion. They could not compare the input and output of steam engines in *physical* terms since they lacked the necessary conceptual tools; the notions of heat, energy and mechanical work still had to be developed. Instead they managed to measure the efficiency of steam engines with the help of parameters which were directly relevant for the practical exploitation of these engines. They measured the input in terms of the amount of fuel consumed, that is in bushels of coal, and the output by the amount of water raised over a certain height, that is foot-pounds. The efficiency, called the *effect* or *duty*, was given by the ratio between output and input in units of foot-pounds per bushel of coal. They used this *techno-economic* notion of efficiency for the comparison of the performance of steam engines.[24]

Due to many improvements, the efficiency of the Newcomen steam engines increased about sixfold during the 18th century, from about 5 million foot-pounds per bushel of coal to about 30 million.[25] One of the main heuristic principles in the attempts to improve steam engine performance was to *save steam* in order to reduce the consumption of fuel.[26] Obviously, steam could be saved by avoiding losses through leakages in the boiler, pipes, valves and cylinder. But engineers, especially Smeaton (1724-1792),[27] discovered much more ingenious ways to lower the consumption of steam by Newcomen engines. Let us briefly consider a few improvements due to this heuristic principle.

Smeaton covered the inside surface of the piston with elm or beech wood because "wood is a much more imperfect conductor of heat than metals."[28] This had a double effect on the performance of steam engines. In the first place, it would save steam during the inlet stroke, because less steam would condense at the cold surface of the piston. Secondly, the cold injection water would be less heated when it got in contact with the inside surface of the piston; therefore less injection water would be needed to create a vacuum.

Smeaton also observed that it was very important to find the right proportion between the surface of the cylinder and its volume.[29] This proportion, which is determined by the ratio between the diameter of the cylinder and the length of the stroke, greatly influenced the amount of steam that was lost in heating the cylinder during the inlet stroke. He had built an engine with a long cylinder of small diameter, but it performed badly. He realized that, because of the great surface of the cylinder, a lot of steam was lost each time the cylinder was heated. The dimensions of an engine had, therefore, to be chosen such that, for a given volume of the cylinder its surface would be as small as possible. Then the loss of steam would be minimal. Moreover, the loss of steam would be relatively less in larger cylinders.

The consumption of steam could also be decreased by adding some air to the steam entering the cylinder.[30] During the heating of the cylinder, steam mixed with air would condense at the surface of the cylinder. As a result, the air would accumulate at that surface. The effect of the air cushion between the surface of the cylinder and the steam inside it would be a diminution of the condensation of steam on the internal surface of the cylinder, and thus a saving of steam.

Although the above improvements greatly increased the efficiency of Newcomen engines, a lot of steam was nevertheless wasted each stroke because of the consecutive heating and cooling of the cylinder. Measurements conducted by Smeaton showed that about three-quarters of the steam was lost for heating the cylinder.[31] Watt (1736-1819) succeeded in avoiding this loss of steam by introducing a separate condenser. This was a fundamental new design principle that revolutionized steam technology. Watt further decreased the consumption of steam by providing the cylinder with a steam jacket and by his principle of expansive action. By cutting off the supply of steam to the cylinder long before the end of the working stroke, steam could be used in a more efficient way.[32]

From the point of view of the principle of saving steam, Watt's improvements had brought the steam engine to a state of high perfection at the beginning of the 19th century. The introduction, at that time, of high pressure steam engines marked the beginning of a new era in which other heuristic principles started to guide attempts to raise the efficiency of steam engines. The accent was put more on the increase of the pressure of steam than on the saving of steam. But as we remarked above, the reliability of this heuristic principle was rather controversial. Carnot's analysis of heat engines played a crucial role in clarifying this matter. It proved of fundamental importance for the further development of steam power technology because it offered new reliable heuristic principles for increasing the efficiency of steam engines. Contrary to the "save steam" principle, these new principles were based on a thoroughly theoretical analysis of heat engines. Let us first have a closer look at the nature of Carnot's work before turning to these new heuristic principles.[33]

The steam engine is omnipresent in the *Reflexions*[34] and according to Fox the book was primarily intended for steam engine engineers.[35] Nevertheless, the problem treated by Carnot is, in my opinion, not a *technological* problem, but a *scientific* one. Carnot was not preoccupied with the analysis of a particular type of artifact, but with a physical phenomenon, namely the *motive power of fire*. The following description by Carnot of the aim of the *Reflexions* makes this clear.[36]

The phenomenon of the production of motion by heat has not been treated from a sufficiently general point of view. It has been treated almost exclusively with respect to engines whose character and operation make it impossible for the full potential of the phenomenon to be realized. In such engines, the production of motion is, as it were, curtailed and incomplete, and it becomes difficult to perceive the principles underlying the process and to study its laws. In order to grasp in a completely general way the principle governing the production of motion by heat, it is necessary to consider the problem independently of any mechanism or any particular working substance. Arguments have to be established that apply not only to steam engines but also to any conceivable heat engine, whatever working substance is used and whatever operations this working substance is made to perform.

Carnot is interested in the *laws* governing the production of motion (work) by heat "independently of any mechanism or any particular working substance." The problem posed by Carnot is itself not a strictly *technological* problem, although it clearly is a problem which arises from the technological context of steam engines. In that context the question as to how the input and output of steam engines are related to each other is of central importance.[37] But Carnot interprets this question in such a way that it is separated from its concrete technological setting. By doing so, he leaves the domain of technology and enters the field of physics. According to Krug, he creates "a model of the ideal technological process at a level of abstraction at which construction parameters still play no role."[38] This he does by characterizing the input and output of steam engines in terms of *physical* quantities. Consequently, Carnot's theory applies to any kind of heat engine, that is any kind of physical system converting heat into mechanical work. Instead of a theory about a particular class of technological artifacts, Carnot, starting from a technological problem, develops the beginning of a whole new branch of physics.

For our purposes, it is not necessary to go into the details of the argument by which Carnot shows that there is a maximum to the efficiency of a heat engine and that this maximum is determined by the temperatures of the heat reservoirs involved. Instead we will focus on Carnot's analysis of the bearing of his theory on the design and construction of steam engines.[39]

To begin with, Carnot clearly points out the conditions under which heat will produce the maximum motive effect:[40]

Hence the necessary condition for the achievement of maximum effect is *that the bodies used to produce motive power should undergo no change in temperature that is not due to a change in volume.* Conversely, whenever this condition is fulfilled, the effect produced will be a maximum.

He immediately stresses the fundamental heuristic value of this principle for steam technology:[41]

This principle must constantly be borne in mind in the construction of steam engines: it is of fundamental importance. If the principle cannot be strictly observed, any departure from it must be reduced to a minimum.

Of course, in practice it will never be possible to obey this principle, that is, to realize an ideal Carnot process. But it clearly indicates the way the efficiency of steam engines can be improved.

A second important conclusion drawn by Carnot is that in principle it does not matter what kind of substance is being used in heat engines to transport heat between the two reservoirs:[42]

The motive power of heat is independent of the working substances that are used to develop it. The quantity is determined exclusively by the temperatures of the bodies between which, at the end of the process, the passage of caloric has taken place.

This is a valuable heuristic conclusion, since it implies that, in principle, any attempt to improve the efficiency of steam engines by using other working substances is in vain; it is no use to go in that direction. For practical reasons, however, certain substances may be preferable above others. According to Carnot, the use of solids or liquids would entail innumerable practical difficulties.[43] Likewise, the use of other gases or vapors instead of steam, such as alcohol, would present practical difficulties.[44] For pragmatic reasons, Carnot recommends the use of steam or air (in the air engine).

Apart from these two fundamental principles, Carnot's theory also clarifies the relation between temperature and the efficiency of steam engines:[45]

It is easy to imagine a host of engines suitable for developing the motive power of heat through the use of elastic fluids. But, whatever approach is adopted, we must not lose sight of the following principles:

1. The temperature of the fluid must first be raised as high as possible, in order to secure a large fall of caloric and thereby the production of a great amount of motive power.
2. For the same reason, the cooling must be carried as far as possible.
3. We must see that the passage of the elastic fluid from the highest to the lowest temperature is brought about by an increase of volume. In other words, we must see to it that the cooling of the gas is a spontaneous consequence of rarefaction.

Again, these principles are important guidelines for designing steam engines. As Carnot points out, practice of course limits the temperatures to which working substances can be raised and the degree of cooling that can be achieved by adiabatic expansion. In the case of steam, high temperatures go hand in hand with high pressures, thus strong vessels and strong adiabatic cooling requires expansion to huge volumes, that is, requires huge cylinders. The first principle also elucidates the superiority of high pressure steam engines above low pressure engines. For saturated steam (all engines at that time were operating with saturated steam) higher pressure means higher temperature, and thus high pressure engines can exploit a greater "fall of caloric," that is, the temperature difference between boiler and condenser is greater than for low pressure engines. Therefore, they are in principle more efficient. According to Carnot:[46]

The mark of a good steam engine, therefore, must be not only that it uses steam at a high pressure *but also that it uses it at pressures which are not constant but which vary substantially from one moment to the next and progressively decrease.*

The pressure has to vary because the cooling of the steam has to be achieved by adiabatic expansion. Carnot describes this principle, which he attributes to Clément, as "the very foundation of the theory of steam engines."[47]

At the end of the *Reflexions*, Carnot compares the efficiency of existing engines with their theoretical maximum efficiency. The figures are disappointing: the efficiency of the best engines known is only about one-twentieth of their theoretical maximum efficiency. So there is still ample room for improvement. But, Carnot warns in the final paragraph, the thermal efficiency is only one criterion to judge the performance of steam engines:[48]

> The saving of fuel is only one of the conditions to be fulfilled in a heat engine. In many circumstances, it is of only secondary importance. It must frequently have a lower priority than the safety, solidity, or durability of the engine, or the constraints on available space, the need to reduce the cost of erection, etc.

In these words, Carnot shows an acute awareness of the limited value of his theory for the practice of steam engine technology.[49] It offers powerful heuristic guidelines for improving steam engines with regard to just one criterion which in practice might not even be the most important one.

Looking at Carnot's work from the perspective of the science-technology issue, we may observe that it does not fit at all into the technology-is-applied-science model. There is no straightforward, direct path leading from Carnot's theory to new inventions in steam power technology. The theory did not induce spectacular new designs of steam engines comparable to Watt's separate condenser, that might be considered to be "applications" of this theory. The ideal Carnot cycle with its maximum efficiency was not embodied in a completely new type of steam engine. As Carnot himself remarked, it was not his "aim here to enter into the constructional details of steam engines."[50] Instead, Carnot's theory specifies "merely" a number of important principles that engineers have to take into account when designing and constructing steam engines. The theory does not spell out how, in practice, these principles can best be fulfilled. Here, engineers had to rely on their technological experience and creativity.

The relevance of Carnot's theory for the development of steam power technology lies mainly in the fact that it provided steam engine engineers with a physical understanding of what was going on in steam engines. Carnot's work solved the controversy about high pressure steam engines; he discovered the real reason why this type of engine was (in principle) more efficient. This reason was not the use of high pressure steam, as most engineers thought, but the use of steam at higher temperatures. He also showed that from a theoretical point of view nothing was to be gained by choosing a particular working medium, but that this choice had to be made on practical grounds.

Carnot's theoretical insights formed the basis for a very powerful heuristic for improving the thermal efficiency of steam engines. He clearly pointed out the ideal conditions under which maximum efficiency could be obtained and how this maximum efficiency could be improved by increasing the temperature difference between boiler and condenser. Contrary to earlier heuristics, which originated mainly from close experience with existing steam engines, this heuristic had a strong theoretical/scientific basis. Compared to the "save steam" or "save fuel" heuristics, Carnot's is much more abstract and general. This is due to the fact that Carnot derives his heuristic principles not from a study of a certain type of technological artifact, but from an analysis of a "theoretical construct," namely a "heat engine," that is, any kind of process transforming heat into mechanical work. Steam engines are just a particular kind of heat engine and therefore his heuristic principles apply also to steam engines. The scientific foundation of Carnot's heuristic principles shows itself precisely in the fact that these principles are not bound to a particular kind of technological artifact.

3. DISCUSSION

Within the context of the science-technology issue, the role of science as a source of powerful heuristics for the solution of technological problems has not received systematic attention.[51] In my opinion, the above examples demonstrate that the heuristics model may be a fruitful framework for dealing with an important aspect of the influence of science on technology.[52] A more detailed conceptual framework for analyzing the heuristic impact of science on technology still has to be developed. Let me put forward here just a few suggestions for further research.

It remains to be seen whether more formal approaches to heuristics, developed within the context of the theory of human problem solving and decision making theory,[53] are of much relevance within the present context. In these theories a heuristic search method is usually interpreted as a set of rules that restrict the search for a solution of a problem to certain parts of a "solution space" and provide "valuations" for possible solutions.[54] It seems doubtful that in cases like those discussed above it will be possible to supply a concrete content to notions like solution space and valuations. Bradshaw has put forward the interesting suggestion that for technological problems the solution space in which the search is conducted may be interpreted either as a design space or a function space.[55]

A more promising approach, in my opinion, would be to try to establish, from a heuristic point of view, a kind of taxonomy of technological problems. Consider, for instance, the following two technological problems: (A) How is it possible to avoid loss of steam in Newcomen engines due to the heating and cooling of the cylinder?, and (B) What is the most favorable point to cut off the steam in a Watt-engine operating expansively? Both problems stem from a desire to improve the efficiency of steam engines but are nevertheless different in nature.

Problem B is an *optimalization problem* in the context of the use of a given steam engine. The set of possible solutions is precisely known: the cutoff point lies somewhere between the beginning and end of the working stroke. The question is to pick out the right solution from this set. This is a highly *structured* problem. It can be solved in principle by a systematic empirical search procedure, but steam engine engineers also developed mathematical/theoretical methods for solving this optimalization problem.[56] Problem A, on the contrary, is a much more open, *unstructured* problem. The set of possible solutions is not known beforehand, but instead has to be generated. Moreover, from the problem itself it is not possible to derive a procedure for generating this set. Watt solved this problem by introducing *a new design principle*; he altered the construction of the Newcomen engine such that one of the functions of the cylinder was taken over by a distinct device, namely the separate condenser. This kind of problem might be termed a *design problem*.

Working along this line of thought, other kinds of problems, encountered in the design, construction and (optimal) use of technological artifacts might be distinguished. It is to be expected that these various types of problems involve different kinds of heuristics. From the point of view of the interaction between science and technology, it would be very interesting to compare the impact of science on the heuristics employed in the solution of various kinds of technological problems.

From a historical point of view, finally, the study of the different origins of heuristic rules and principles deserves closer attention.

In this respect, I have in mind primarily the opposition between science-based and experience-based heuristics.[57] The 18th century "save steam" heuristic is strongly experience-based and is typically phrased in a technological language, that is, employs concepts referring to (parts of) technological artifacts. The Carnot heuristic, on the contrary, is phrased in a rather abstract, scientific language. Historically, the emergence of science-based heuristics is of utmost importance; it marks the beginning of a growing impact of science on the development of steam technology. Generally speaking, the study of the emergence of science-based heuristics may prove a fruitful way for tracing the growing influence of science on technology.

University of Technology – Eindhoven

NOTES

1 For a discussion of the "technology is applied science" view, see Staudenmaier (1985, p. 95 ff).

2 See Staudenmaier (1985), chapter 3.

3 For the various meanings of the notion "technological determinism," see Chant (1989), sect. 2.2.

4 See Van den Belt & Rip (1984), p. 37.

5 See Van den Belt & Rip (1984), p. 48; following a suggestion made by Nelson and Winter, they also emphasize the heuristic influence of science on technology.

6 This process was a major technological breakthrough; ammonia was, among other things, a crucial raw material for the production of fertilizers and explosives.

7 This section is based on Kroes (1985).

8 For more details about the Heat Theorem, see Nernst (1969).

9 In 1920, Nernst was awarded the Nobel Prize in chemistry for his Heat Theorem.

10 For more details about how this was done, see Nernst (1969).

11 The following, schematic overview of the history of the synthesizing of ammonia is based on Mittasch (1951).

12 Produced by BASF, the firm for which Haber was working after 1908.

13 For more details, see Mittasch (1951).

14 Mittasch (1951, p. 72); translation by the author.

15 Mittasch (1951, p. 63 ff.).

16 Mittasch (1951, p. 12).

[17] A similar situation occurred with regard to the search for an antiknock additive for gasoline; huge numbers of compounds were tried. See Hughes (1989, pp. 223-4).

[18] Mittasch (1951, p. 75); translation by the author.

[19] For a description of the Woolf engine, see Hills (1989) and Farey (1971), chapter XI.

[20] See Fox (1976); see also Payen (1976).

[21] For a description of various attempts to replace steam, see Babcock (1886).

[22] This problem was studied extensively in France between 1815 and 1824, the year of the appearance of Carnot's *Reflexions*, by people like Clément, Desormes, and Navier; see Fox (1976).

[23] I will use here the version translated and edited by Fox; see Fox (1986).

[24] For more details about this techno-economic notion of efficiency, see Kroes (1991).

[25] See Cardwell & Hills (1976), p. 1.

[26] The saving of fuel could also be achieved by other means, for instance, by improving the quality of the pumps (see Hills (1989), p. 101), by feeding the boiler with warm water from the top of the cylinder or the heated injection of water (see Desagulier (1751), p. 102), or by the construction of better boilers (see Smith (1980), p. 207-208). Here, however, we will focus our attention on attempts to save steam.

[27] For Smeaton's work on steam engines, see Allen (1981).

[28] Quoted in Farey (1971), vol. 1, p. 251.

[29] See Rees (1819), pp. 86-90.

[30] See Farey (1971), p. 170.

[31] See Pacey (1974), p. 136.

[32] See Hills (1989), pp. 64-5.

[33] For a more extensive discussion of the nature of Carnot's work, see Kroes (1992).

[34] See Payen (1976), p. 125.

[35] Fox (1986, p. 2).

[36] Fox (1986, p. 63-4).

[37] According to Fox, the background of Carnot's work is the debate about the economy of the Woolf engine [Carnot, 1986, p. 8]: "It is true that Carnot constructed his theory without reference to any particular engine, but I am convinced, none the less, that his work should be seen as a contribution to the lively debate, centering on the question of economy and

conducted almost exclusively in France, that was sparked off by the recognition of the remarkable qualities of the Woolf engine in about 1815."

38 Krug (1981), p. 8; translation by the author.

39 See particularly the final pages of the *Reflexions*: Fox (1986), pp. 102-113.

40 Fox (1986, p. 70).

41 *Ibid.*

42 Fox (1986, p. 76-7); italics in the original.

43 See Fox (1986, p. 102).

44 See Fox (1986, pp. 108-111).

45 Fox (1986, p. 102-3).

46 Fox (1986, p. 104).

47 *Ibid.*

48 Fox (1986, p. 113).

49 See also the commentary by Fox on Carnot's text, Fox (1986), p. 153.

50 Fox (1986, p. 108).

51 See, for instance, Staudenmaier's survey of various views on the interaction between science and technology in Staudenmaier (1985), chapter 3.

52 Studies by Van den Belt & Rip (1984), Vermeij (1989) and Hutter (1988, chapter IX) also underline the importance of science-based heuristics for the development of technology. Note that effective heuristics may be derived from scientific ideas that later on turn out to be wrong; for an example, see Kline (1987).

53 See Newell & Simon (1972), chapter 4.

54 See Winograd & Flores (1986), p. 22.

55 For more details, see Bradshaw [1992]. I owe this reference to an anonymous referee.

56 See, for instance, Pambour (1839).

57 An important source for heuristics in technological problem solving is analogies and metaphors. For a classification of different kinds of analogies in technology, see Sarlemijn & Kroes (1988). Especially, analogies with already existing technological devices may generate heuristic rules. For an example see Sarlemijn (1987) which contains a discussion of the different kinds of analogies involved in transistor research.

REFERENCES

Allen, J. S. (1981). "Steam Engines," in A.W. Skempton (ed.), *John Smeaton, FRS*. London: Thomas Telford Limited, pp. 179-194.

Babcock, G.H. (1886). "Substitutes for Steam." *Transactions of the American Society of Mechanical Engineers*, 7: 680-741.

Belt, H. van den & A. Rip (1984). *Technologie-ontwikkeling: het Nelson-Winter/Dosi-model*. Leiden: Lisbon/R-84/21.

Bradshaw, G. (1992). "The Airplane and the Logic of Invention," in R. N. Giere (ed.), *Cognitive Models of Science*, Minnesota Studies in the Philosophy of Science, vol. XV. Univ. of Minnesota Press, Minneapolis.

Cardwell, D. S. L. & R. L. Hills (1976). "Thermodynamics and Practical Engineering in the Nineteenth Century," in A. R. Hall & N. Smith (eds.), *History of Technology*. London: Mansell.

Chant, C. (ed.) (1989). *Science, Technology and Everyday Life 1870-1950*. London: Routledge.

Desaguliers, J. T. (1751). *De natuurkunde uit ondervindingen opgemaakt*, vol. 3. Amsterdam: Isaak Tirion. (Dutch translation of *A course of Experimental Philosophy*.)

Farey, J. (1971). *A Treatise on the Steam Engine*, 2 vols. Newton Abbot: David & Charles Reprints. (Siginally published 1827.)

Fox, R. (1976). "The Challenge of a New Technology: Theorists and the High-Pressure Steam Engine Before 1824," in *Sadi Carnot et l'essor de la thermodynamique*. Paris: Éditions du Centre National de la Recherche Scientifique, 1976, pp. 149-168.

Fox, R. (ed.) (1986). *Sadi Carnot: Reflexions on the Motive Power of Fire*. Manchester: Manchester Univ. Press.

Hills, R. L. (1989). *Power From Steam*. Cambridge: Cambridge Univ. Press.

Hughes, T. P. (1989). *American Genesis*. New York: Penguin Books.

Hutter, J. J. (1988). *Toepassingsgericht onderzoek in de industrie; de ontwikkeling van kwikdamplampen bij Philips 1900-1940*, Ph.D. Thesis. Eindhoven: University of Technology.

Kline, R. [1987]. "Science and Engineering Theory in the Invention and Development of the Induction Motor, 1880-1900." *Technology and Culture*, vol. 28, pp. 283-313.

Kroes, P. A. (1985). "De technische relevantie van fundamenteel onderzoek; het Warmte-theorema van Nernst en de ammoniaksynthese," in A. Sarlemijn (ed.), *Van natuurfilosofie tot technische natuurkunde*. Eindhoven: EUT report, Department of Physics, Eindhoven University of Technology, pp. 32-40.

Kroes, P. A. (1991). "Steam Engines and the Concept of Efficiency; Characteristics of Technological Knowledge." *Methodology and Science*, vol. 24, no. 2, pp. 79-97.

Kroes, P. A. (1992). "On the Role of Design in Engineering Theories; Pambour's Theory of the Steam Engine," in P. A. Kroes & M. Bakker (eds.), *Technological Development and Science in the Industrial Age*. Dordrecht: Kluwer, pp. 69-98.

Krug, K. (1981). "Zur Herausbildung der Technischen Thermodynamik am Beispiel der Wissenschaftlichen Schule von G. A. Zeuner." *NTM-Schriftenreihe Geschichte der Naturwissenschaften, Technik und Medizin*, 18, no. 2, Leipzig, pp. 79-97.

Mendelssohn, K. (1973). *The World of Walter Nernst*. London: Macmillan.

Mittasch, A. (1951). *Geschichte der Ammoniaksynthese*. Weinheim: Verlag Chemie.

Nernst, W. (1969). *The New Heat Theorem*. New York: Dover. (Originally published in German in 1917.)

Newell, A. & H. A. Simon (1972). *Human Problem Solving*. New York: Prentice Hall.

Pacey, A. J. (1974). "Some Early Heat Engine Concepts and the Conservation of Heat." *British Journal for the History of Science*, 7, no. 26, pp. 135-145.

Pambour, G. de (1839). *The Theory of the Steam Engine*. London: John Weale.

Payen, J. (1976). "La Pratique des machines a vapeur au temps de Carnot," in *Sadi Carnot et l'essor de la thermodynamique*. Paris: Éditions du Centre National de la Recherche Scientifique, 1976, pp. 125-148.

Rees, A. (1819). *The Cyclopedia, or Universal Dictionary of Arts, and Literature*. London.

Sarlemijn, A. (1987). "Analogy Analysis and Transistor Research." *Methodology and Science*, 20 (3), pp. 40-61.

Sarlemijn, A. & P. A. Kroes. "Technological Analogies and Their Logical Nature," in P. T. Durbin (ed.), *Technology and Contemporary Life*. Dordrecht: Reidel, pp. 237-255.

Smith, A. (1980). "The Newcomen Engine at Passy, France, in 1725: A Transfer of Technology Which Did Not Take Place." *Transactions of the Newcomen Society*, 50, pp. 205-217.

Staudenmaier, J. M. (1985). *Technology's Storytellers*. Cambridge, Mass.: MIT Press.

Vermeij, E. (1989). *Contextuele verschillen in de ontwikkeling van technische toepassingen van methaangisting*. Master's thesis. Eindhoven: University of Technology, Dept. of Philosophy and Social Sciences.

Winograd, T. & F. Flores (1986). *Understanding Computers and Cognition*, Reading, Mass.: Addison-Wesley.

EDWARD WACHTEL

TO AN EYE IN A FIXED POSITION: GLASS, ART AND VISION

In this article, I will discuss a technology which I believe has had a crucial influence on Western culture: transparent glass. Clear glass was the material basis for a number of significant inventions from the thirteenth to the seventeenth century, notably eyeglasses, the microscope and the telescope. Since preliminary investigations of these tools have been undertaken–particularly with regard to their role in exploring the microcosm and the macrocosm–I will not discuss them here. (See, for example, Mumford 1934, pp. 124-131; de Solla Price 1984, pp. 106-110; and Ihde 1990, pp. 52-58.)[1]

I will consider one glass-based technology which has gone all but unnoticed by historians and other students of technology. While I can't be certain, I suspect the reason for this scholarly neglect has to do with one of the physical characteristics of this technology–it is almost invisible. The technology is the transparent glass window, and I will argue that it provided a new way of seeing the world–first for artists, and later for scientists–which fostered, in both art and science, what we call the Western world view.

Since the phrase, "Western world view," means different things to different people, I will briefly explain what I mean by it. In the West–and only in the West–a particular and peculiar view of physical existence has predominated. It has to do with the way we organize space and time. As best I can describe this view, we believe that space and time are separate entities: that the space we inhabit is rather like an empty cardboard box (without the cardboard), which flows along a constant current of time. Or, as Albert Einstein (1961, p. 150) once put it, we think of physical reality as a three-dimensional space which evolves in one-dimensional time.

This view of space and time is visible in Western painting. It is created by the compositional structure which we call perspective. Before I describe the influence of the glazed window, I must take a few minutes to explore perspective as a mode of perception, a style of representation and a way of organizing space and time.

Joseph C. Pitt (ed.), New Directions in the Philosophy of Technology, 41-61.

1. PERCEPTION AND SPACE/TIME

It must be emphasized that perspective in painting is very much a result of a trained *perceptual* technique, not simply a representational style. This point is rarely appreciated by art historians. Even Erwin Panofsky ([1927] 1991), who understood perspective as symbolic form, who comprehended the subtle connections between representation and philosophical change, did not explore, in sufficient depth, the more fundamental relations of visual perception to space conception.

Jean Piaget (Piaget & Inhelder [1948] 1967) has described the long journey from the blooming confusion of infancy to a mature, visual world of stable spatial and temporal relations. He emphasized that we experience spatiotemporal reality through interactions of our sensory and motor apparati. For purposes of this discussion, I will simplify this to say, we build our conceptions of space/time through vision and movement.

But are space/time relations constant from individual to individual, or from culture to culture? The results of many cross-cultural studies suggest that there are different ways to organize the visual world.[2] I suggest that different organizations of the world, different images of space/time result from different modes of perception.

I use the phrase "modes of perception" to distinguish my approach from the approach and procedures commonly used in studies of visual perception. Most studies fix the eye of the beholder and the object(s) beheld. This procedure results in similar or identical retinal images for all subjects. Unremarkably, these subjects tend to see the same thing. However, when researchers have purposely (or unwittingly) left the viewing conditions unspecified, substantial perceptual differences do emerge.

For example, in the mid 1950s, Gordon Allport and Thomas Pettigrew (1957) used an illusion-producing device–the Ames rotating trapezoidal window–to test for perceptual differences among rural South African Zulus, urban Zulus and Western whites. Under optimal, illusion-producing conditions (fixed, single-eye viewing from twenty feet), Allport and Pettigrew found no significant differences among the groups. Almost all subjects experienced the illusion of an oscillating rectangle. Under sub-optimal conditions (binocular viewing without head restraint from ten feet), the rural Zulus saw a rotating trapezoid–the actual shape and movement–while a majority of urban Zulus and whites continued to experience the illusion.

Neither Allport and Pettigrew nor subsequent researchers could adequately account for both sets of results. Allport and Pettigrew chose to emphasize their results under optimal conditions, and reported that their experiment offered no grounds to conclude that environment influenced perception (1957, p. 111). After reviewing the same data, Segall, Campbell and Herskovits came to the opposite conclusion: they found that "there are genuine perceptual differences that are determined by environmental factors" (1966, p. 67).

The experiment did uncover perceptual differences as Segall *et al.* claimed, but to adequately explain both sets of results, we have to examine the viewing conditions, the modes of perception. Briefly, prior researchers have assumed that sub-optimal viewing implied a single condition for all subjects. Rather, sub-optimal viewing provided subjects from each group the (limited) freedom to choose their viewing conditions. The majority of rural Zulus chose to utilize head movement and possibly to switch lead eyes. They chose to see the demonstration from multiple perspectives, while the other groups continued to use a single, fixed-eye mode of perception.

These two modes of perception result in different ways of organizing the world and in different conceptions of space/time. In the next section, I will attempt to link modes of perception to representation and space conception with greater precision.

2. PERSPECTIVE, REPRESENTATION AND SPACE/TIME

All painters confront the same problem when they approach their canvases: how to represent the four dimensions of visual experience on a two-dimensional surface. Daniel Alain, in a 1955 cartoon, addressed the problem with humor (Figure 1). Ernst Gombrich also began his wonderful book, *Art and Illusion* (1960) with this image. Both Alain and Gombrich ask, why did different cultures represent the world in such different ways?

Alain created his "joke" by using perspective composition to represent the space which his characters inhabit. Yet he places the model in a position we recognize as typical of Egyptian composition. The incongruity presents a forceful and funny reminder that the Egyptian method of representation looks flat or distorted to our eyes.

Figure 1.

Perhaps a closer look at how the Egyptians organized their world in paint may shed some new light on perception, representation and cultural differences in the organization of space and time.

Most analysts of perspective and perception–among them, Gombrich (1960), Panofsky (1927) and White (1967)–have tended to focus on the *spatial* result of choosing one or multiple perspectives (i.e., "flat" space results from the artist choosing multiple viewpoints, and deep or three dimensional space is concomitant with single, fixed eye viewing). My own focus, on the *temporal* implications of viewpoint choice, is less common but equally valid, and for purposes of this essay, more generative. For some compelling arguments for the necessity of including time and motion in any analysis of perspective and vision, see Goodman (1976, pp. 10-19) and Piaget (Piaget and Inhelder 1948, pp. 13-17).

Figure 2 is a fresco from the tomb of Cha, and it is typical of the Egyptian solution to the problem of spatiotemporal representation. (Although the earliest examples of this form of representation precede Egyptian civilization by perhaps a millennium, I will continue to refer to this method as the "Egyptian" solution.)

Figure 2

While our first response to this image may be to notice the flatness of the space, we can view the flatness as a result of the way Egyptian painters represented *time*. The time that is represented in this painting is the time it would take to see the subjects from the different viewpoints represented. For example, imagine for a moment that you are in the presence of Cha and his wife. Take a look at Cha's face. We are offered a side view. We would have to move to see his chest from the front, move again to see the side view of his legs, again to see the front of the chair, etc. Therefore, what the Egyptian painters captured, through multiple viewpoints, was visual experience *over time*. In the twentieth century, many painters returned to the Egyptian method to capture visual experience in time. The early cubist works of Piccasso and Braque may represent the most direct applications of the multiple viewpoint to the problem of organizing experience in two dimensions.

Although art historians attribute a number of technical and aesthetic advances to the Greeks, the available evidence confirms that they continued to represent the temporal dimension just as the Egyptians did. For example, Figure 3, a vase painted in the black figure style, demonstrates a more complex use of overlapping forms and a greater fluidity than Egyptian painting exhibits. Yet we are still presented with the multiple viewpoints and multiple times of the Egyptians. Compare the central figure on this vase with Cha in Figure 2. In both, we see a side view of the head, front view of the chest, side view of the legs, etc.[3]

Figure 3

From the earliest civilizations through Greek art–perhaps five thousand years–there were no fundamental changes in the representation of time and space in painting. Finally, in the first century B.C., a sudden revolution occurred in the representation of visual experience. On Roman walls, in the resort towns of Pompeii and Herculaneum, a new way of seeing was recorded. While the contents and motifs of these frescos and mosaics were often borrowed from Greek models, they displayed a spatial construction "never seen before and not to be seen again until the Italian Renaissance, nearly fourteen or fifteen centuries later" (Bornstein 1961, p. 55).

The illusion of depth in these images is created by viewing the world with a single eye from one, fixed position. The slaves who painted these pictures no longer permitted themselves nor the viewer the freedom to move and to see from multiple viewpoints. The freedom to know subjects from different angles and from different times was denied. In short, they excluded time from visual experience. While the Romans rarely, if ever, achieved the mathematical purity of Renaissance perspective, they clearly developed the perceptual technique on which it is based: the fixed viewpoint.[4]

I have included a Roman still life to illustrate their achievement (Figure 4). In this picture, we can see not only the illusion of depth created by restricting the viewpoint, but also the perspectival "imperfections" which result from not fixing it completely. The table, cup, platter and pitcher all appear three dimensional, but the cup and platter seem to be tilted forward in relation to the table. The tilt is the result of painting the cup and platter from a slightly higher viewpoint. (This fresco is somewhat reminiscent of some of Cezanne's still lifes. Perhaps Cezanne and the Roman painter were engaged in similar struggles with perspective: the Roman struggled to achieve it; Cezanne, to overcome it.)

When the Roman Empire fell, fixed space and time disappeared from painting. Throughout the middle ages, the only vestiges of the Roman advances are foreshortened subjects and frontal settings. Time, in the form of multiple viewpoints, returned to painting and remained a part of visual experience for almost a millennium. Figure 5 may illustrate the organization of space and time typical of medieval painting. Notice how Luke's gospels are turned toward us; and how the footstool is also presented from a different viewpoint.

Figure 4

With the Italian Renaissance comes another revolution in representation. Perspective suddenly reappears, first on Italian walls and soon, throughout Europe.

From an example of pure perspective construction, we can begin to understand the visual conditions which are represented. In Raphael's "Marriage of the Virgin" (Figure 6), the parallel lines of the plaza all meet at the vanishing point. Raphael was kind enough to paint an open doorway in the dome so we can see the horizon where they meet. You should be able to feel how this construction forces you to view the image from one fixed place. (As I write this, I cannot be sure whether the picture will be reproduced large enough, and in sufficient detail, to feel the correct viewing position. If not, try any Renaissance picture with strong perspective lines in an art book, or better, in a museum.) Further, "correct" viewing demands that only a single eye be used. By fixing the viewing conditions in this way, perspective representation freezes all movement between the beholder and the subject.

In describing the effect of perspective on time, Jacob Bronowski once said, "not so much a place as a moment is fixed for us, and a fleeting moment: a point of view in time more than in space" (1973 p. 78). I will take the matter a bit further: in perspective representation, time is reduced not to the moment, but to the non-existence of a theoretical instant, and the remaining three dimensions of existence are envisioned as a rigid box of space.

Figure 5

How is the window related to the appearance of perspective in Rome, its disappearance during the middle ages and reappearance in the Renaissance? Historically, the relationship is a close one. The first manifestations of the perspective conception of space were the wall paintings on Roman villas. The glazed window, it seems, was also a Roman invention. (The Greeks had no windows, or even window openings in their houses.)

Figure 6

3. HISTORY OF THE WINDOW

At first, the Romans used *lapis speculoris*, a substance like mica for their windows. This material was split into translucent sheets and cemented into the window holes (Rogers & Beard 1937, p. 138). Later, the Romans learned to roll molten glass over flat stones to make window glass that was rather lumpy and not quite transparent (Logan 1951, p. 14). By the second century B.C., methods of glass making had improved to the point where glass was quite transparent, although the sheets were small (less than 12" x 12") and slightly tinged with green or blue (Sheppard & Wright 1948, p. 12). During the first century B.C., casting and pressing methods were discarded when the Romans developed glass blowing (Ellis

1993, p. 62). Transparent glass became relatively cheap and common.

When Pompeii was excavated, window panes and their lead frames were discovered along with some of the earliest examples of perspective art.

It seems that the invention and use of transparent windows preceded the birth of perspective by a few generations. Furthermore, both developments occurred in the same culture. So, the historical evidence would support a claim that the window fostered perspective in Rome. But what about the disappearance of perspective in the middle ages? Was the window related to this change in space conception? Once again, the historical evidence suggests that it was.

Window glass was totally forgotten with the fall of the Roman Empire. In the middle ages, the architecture of the nobles took on the aspect of a fortress, while peasant building was represented by wattle and daub huts. For protection, the "windows" of the feudal castle were reduced to slits and raised above eye level. Their main function was to let out the smoke of fires, not to let in a view (Sheppard & Wright 1948, p.13). Meanwhile, the glass industry shifted eastward with Constantinople as its capital. In the seventh century, the Syrians found a more efficient way to produce window glass. By blowing a bubble of glass, then spinning it, the centrifugal force flattened the bubble into an even sheet. Although this glass was sprinkled with bubbles, tinted and translucent rather than transparent, it was the material used for the magnificent stained glass windows which, over the next few centuries, spread to the churches and cathedrals throughout Europe.

By the thirteenth century, the oppression of architecture by the fortress aesthetic had slowly lifted. Windows, instead of being small slits or smoke holes, gradually expanded in size. Concurrently, the Venetians had acquired glass-making skills from their colony in Syria. They developed techniques for making transparent glass and soon became the glaziers for the Italian peninsula, and later, for all of Europe (Logan 1951, pp. 14-16).[5]

The history of the window and the perspective space conception both began in Rome at about the same time, were jointly forgotten during the middle ages, and were re-discovered on the Italian peninsula at the start of the Renaissance.

4. THE WINDOW AND PERSPECTIVE

So much for the historical relationship. I will now turn to the logic of the connection or, more precisely, the question of how the window provided an epistemological tool for training perception.

I will begin with an observation made by Leon Battista Alberti, the man who first codified perspective theory. In 1435 he said, "I consider [my canvas] to be a . . . window through which I view that which will be painted there" (White 1967, p. 122).

Somewhat later, Leonardo Da Vinci concurred. He said, "Perspective is nothing else than the seeing of an object behind a sheet of glass, smooth and quite transparent, on the surface of which all things may be marked that are behind this glass" (MacCurdy 1939, p. 992.).

One of the points implied by these observations is that the surface to be painted should be viewed as a transparent plane between the painter or the viewer, and the reality to be represented. It requires that we negate a material surface and reinterpret it as non-material plane (Panofsky 1927). For most of us, this conception of the picture plane is quite simple to imagine, even "natural." Yet, in a culture without clear glass, it would be difficult to conceptualize this arrangement. How could one even imagine a wall, a canvas or any other opaque material between ourselves and a subject to be transparent? In this sense, the glazed window provided a physical model for the basic conception underlying perspective representation.

Furthermore, glass also served as a tool for perceptual training. The act of looking through a window enforces the rigid single-eyed viewpoint which is necessary for perspective. I think I can explain this point more clearly with some visual aids.

At the end of the 15th century Albrecht Dürer traveled to Venice to learn about perspective. He made a large number of prints to demonstrate the Italian discovery to his northern countrymen. In his "Draughtsman Drawing a Reclining Nude" (Figure 7), Dürer shows how a painter might utilize a window to create a perspective rendering directly on paper. In other prints he shows a technique for drawing directly on a window (which would later be traced onto paper). This latter method was advocated by Leonardo as a way to learn perspective (Linscott 1957, p. 63).

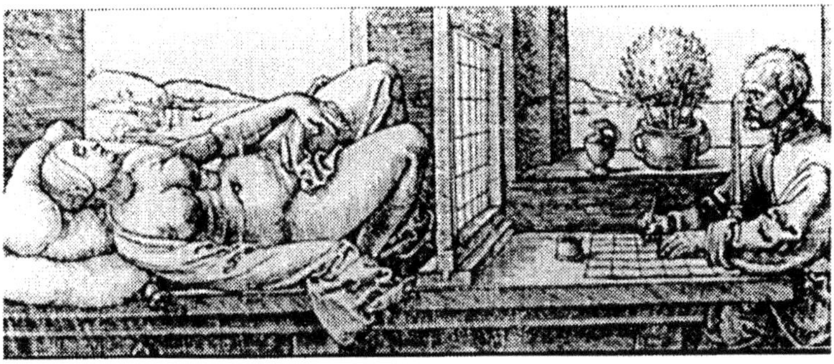

Figure 7

Dürer's print demonstrates the perceptual conditions required to achieve perspective. Note the eye-piece which serves to fix the position of the single eye the painter must use. In fact, the eye-piece is barely necessary. Once two points or a line have been traced on a window, the painter's single eye will be fixed by the correspondence of the line to the reality behind the window. Perhaps the best way to comprehend the relationship of the window to perspective and perception, would be to try it yourself. If you trace the scene outside onto your window, you will be able to make a perfect perspective composition. You should also discover how, to keep the correspondence of sketched lines to outside reality, you are forced to use one eye in a fixed position.[6]

I have presented the historical and theoretical relations between the window and perspective in painting. However, I want to address a possible objection. We tend to think of perspective as "the best" or the only way to represent depth in painting. Perspective construction seems central to what we call "realism." Hence, the art and representations of cultures which haven't developed it are often seen as immature or primitive. There is at least one culture in which the art of painting was highly evolved, which developed its own successful methods of representation and which never utilized glazed windows. I speak of Chinese painting.

Figure 8 is a scroll painting called "Summer Mountains," and it was made by an unknown artist in the 11th century. It presents the style of representing depth which the Chinese developed as early as 200 A.D., and which remained their essential method until recent times.

Figure 8

The correct way to view this painting is to approach it from the right. Your eyes will be guided to something in the foreground, and from that point, up a path or a river into the mountains. Next, your eyes will find another starting point and again be led back into the mountains. I have included two details which illustrate two of the "starting points." In Figure 9, which is cropped from the far left of the larger composition, you will be led from the waterfall upwards to the small house, and then up and back into the mountains. In Figure 10 (from the center of the composition), you will begin with the two figures crossing the bridge in the lower right of the detail. Again, your eyes will be guided up to the village in the mountains.

Unlike Western perspective, the system of representation used by the Chinese demands a moving eye. Although the Chinese methods differ from the Egyptian ones discussed earlier, they share with the Egyptians, with the Greeks and with every non-Western culture, a vision of space permeated with time.[7]

I have focused my remarks on the influence of the window on the development of perspective in Western painting. However, the complete separation of time and space that results from this conception has implications that go far beyond the realm of art. I suggest that the window, and the picture of space and time it presents, may also have influenced the sciences.

Figure 9

In 1686, Sir Isaac Newton published the *Principia*, his mathematical system of the physical world. His system explained the motion of bodies in space, and used a mathematics of time, called the fluxions or calculus. From the viewpoint of twentieth century hindsight, we know that his system was flawed. The fundamental flaw was not essentially mathematical, but conceptual. Newton's bodies moved in a space that was three-dimensional and absolute. His calculus described a time that was linear, separate from space and equally unchangeable. In short, Newton did not describe reality, but a particular, and it so happens, incorrect picture of reality. Where did this picture come from? I submit that the complete separation of space and time which provided the foundation of the mechanical world view had been painted on canvases and cathedrals since the

15th century. By Newton's time, this view of existence had penetrated to the deepest levels of Western consciousness. That is to say, it had become common sense.

Figure 10

I began the discussion of perspective and representation with a visual joke to establish the problem to be addressed. I will also end with one. In one image (Figure 11), René Magritte makes fun of the various relationships I tried to describe. Observe how he establishes the identity of the window view and the canvas by having them melt into one another. Next, he makes fun of the space that results from this mode of perception. He calls the picture "The Promenades of Euclid." Note the plural. He is telling us that the space created by this way of seeing is a fiction and an absurdity. It is a space so preposterous that the upright spire in the foreground has the same form as the road which leads us back to the vanishing point. Both the spire and the road are the promenades of Euclid. The space and time they inhabit are the space and time the West has inhabited since the invention of the window.

Figure 11

Fordham University

ILLUSTRATIONS AND CAPTIONS

1. Daniel Alain, Cartoon, originally published in *The New Yorker* (October 1, 1955).

2. Fresco of Cha and His Wife, from the tomb of Cha (New Kingdom). Egyptian Museum, Cairo.
3. Vase, Black Figure Style, (ca. 525 B.C.). British Museum, London.
4. Still Life with *Eggs and Game*, From the House of Julia Felix, Pompeii (Early 1st Century). Naples Museum.
5. Portrait of St. Luke, From *St. Augustine's Gospels* (Late 6th Century). Corpus Christi College Library, Cambridge, England.
6. Raphael Santi, *Marriage of the Virgin* (ca. 1505). Brera, Milan.
7. Albrecht Dürer, Woodcut. *Draughtsman Drawing a Reclining Nude* (ca. 1527).
8. Anonymous Scroll Painting in the Yen Wen-Kuei Tradition, *Summer Mountains* (ca.1050).
9. Detail of the left side of *Summer Mountains.*
10. Detail of the center of *Summer Mountains.*
11. René Magritte, *The Promenades of Euclid* (1955). Minneapolis Institute of Art.

NOTES

1 See Ihde (1990, ch. 5), for his discussion of transparency and embodiment which directly relates to my thesis.

2 See Deregowski (1972) for a comprehensive overview of cross-cultural studies in pictorial perception up to 1970. It should be noted that studies of pictorial perception have found more regular and dramatic differences in vision than studies which utilize actual (four-dimensional) objects and events. The discussion of the research of Allport and Pettigrew which follows may shed some light on the problems of testing for perceptual differences with real objects and events.

3 Unfortunately, Greek wall painting has entirely disappeared. A few historians have argued for the development of a rudimentary perspective in Greek painting, but from the direct evidence–vase painting–no case can be made for it. See G. Richter (1971) for an enthusiastic case for Greek perspective, J. White (1967) for his careful distinctions between the Roman and Greek contributions, and W. Ivins Jr. (1964), who shares my view that perspective was a Roman discovery. Those who attribute perspective to the Greeks rely rather heavily on Roman sources and Roman paintings which are presumed to be derived from Greek works. They may be confusing derived content for derived form. As an example, every tourist shop in Greece sells imitations of classical vases. Many of them utilize perspective form to represent the figures–Achilles, Heracles, Aphrodite, etc.–which appeared on the original vases in the "Egyptian" form. I suggest that Roman painters did the same thing–represented Greek content in perspective style.

4 Not only did the Romans develop the practical techniques for perspective, they also came close to offering a theory. Vitruvius, in his *De Architectura*, provides all the rudiments for such a theory. See White (1967, pp. 249-262) for an analysis of Vitruvius' contributions.

5 The importance of glass was such that "to guard the industry's trade secrets, the Venetians transported the industries to the Island of Murano . . . . Hand picked armed guards patrolled Murano's 'mile of glass houses' where thousands of glassmakers worked behind iron bars." Glassmakers were raised to the rank of burghers, and "were privileged to marry off their daughters to noblemen" (Logan (1951, p.15).

6 My analysis has been focused on the glass pane rather than the window frame. A number of cultures have used window openings and windows containing translucent materials such as paper, mica, impure glass, etc. While an empty frame can be used to limit the viewpoint, it cannot fix it completely. Further, an empty or translucent window provides no model for the concept which is central to perspective representation: that of a plane to be seen through.

However, to a mind already familiar with the concept of a transparent plane, an open window or doorway may serve as an aid to fix a composition. In fact, the first pure perspective work–"S. Giovanni and the Piazza del Duomo," by Filippo Brunelleschi–was created with the help of a doorway. Although this work has been lost, its construction was documented by Antonio Manetti in his *Life of Brunelleschi*. According to Manetti, Brunelleschi retreated about two meters inside the main door of the Duomo to restrict his viewpoint and limit the field of the piazza (White, p. 28).

7 For related discussions of technological influences on Chinese art, see Wachtel & Lum (1991); on Paleolithic cave painting, see Wachtel (1993); and on the influence of television on contemporary space conception, see Wachtel (1985).

REFERENCES

Allport, Gordon, and Pettigrew, Thomas (1957). "Cultural Influence on the Perception of Movement: The Trapezoidal Illusion Among the Zulus." *Journal of Abnormal and Social Psychology*, 55:104-113.

Bornstein, Eli (1961). "The Window on the Wall." *The Structurist*, 1: 51-57.

Bronowski, Jacob (1973). *The Ascent of Man*. Boston: Little, Brown and Co.

Deregowski, Jan (1972). "Pictorial Perception and Culture." *Scientific American*, 227:82-88.

Einstein, Albert (1961). *Relativity: The Special and General Theory*, 15th ed. New York: Crown Publishers.

Ellis, William S. (1993). "Glass: Capturing the Dance of Light." *National Geographic*. December. Pp. 37-69.

Gombrich, E. H. (1960). *Art and Illusion: A Study in the Psychology of Pictorial Representation.* Princeton, New Jersey: Princeton University Press.

Goodman, Nelson (1976). *Languages of Art.* Indianapolis: Hackett.

Ihde, D. (1990). *Technology and the Lifeworld: From Garden to Earth.* Bloomington: Indiana University Press.

Ivins, W., Jr. (1964). *Art and Geometry: A Study in Space Intuitions.* New York: Dover.

Linscott, R. (ed.) (1957) . *The Notebooks of Leonardo DaVinci.* New York: Modern Library.

Logan, H. (1951). *How Much Do You Know About Glass?* New York: Dodd, Mead.

MacCurdy, E. (ed.) (1939). *The Notebooks of Leonardo DaVinci.* New York: Reynolds and Hitchcock.

Mumford, L. (1934) *Technics and Civilization.* New York: Harcourt Brace & World.

Panofsky. Erwin. [1927] (1991). *Perspective as Symbolic Form.* Translated by Christopher Wood. New York: Zone Books.

Piaget, J., and Inhelder, B. [1948] (1967). *The Child's Conception of Space.* Translated by F. Langdon and J. Lunzer. New York: Norton.

Richter, G. (1971) . *Perspective in Greek and Roman Art.* London: Phaidon.

Rogers, F., & Beard, A. (1937). *5000 Years of Glass.* New York: Frederick A. Stoken.

Segall, M., Campbell, D., and Herskovits, M. (1966). *The Influence of Culture on Visual Perception.* Indianapolis: Bobbs-Merrill.

Sheppard, R., & Wright, H. (1948). *Building for Daylight.* London: George Allen.

de Solla Price, D. (1984). "Notes Toward and Philosophy of the Science/Technology Interaction," in *The Nature of Technological Knowledge: Are Models of Scientific Change Relevant?*, edited by R. Laudan. Dordrecht: Reidel.

Wachtel, E. (1993). "The First Picture Show: Cinematic Aspects of Cave Art." *Leonardo*, 26:135-140.

Wachtel, E. (1985). "The Impact of Television on Space Conception," in *Studies in Mass Communication and Technology,* edited by Sari Thomas. New Jersey: Ablex Publishing. Pp. 168-174,

Wachtel, E., & Lum, C. (1991). "The Influence of Chinese Script on Painting and Poetry." *ETC.: A Review of General Semantics,* 48: 275-291.

White, J. (1967). *The Birth and Rebirth of Pictorial Space*. Boston: Boston Book and Art Shop.

SCOTT D. N. COOK

THE STRUCTURE OF TECHNOLOGICAL REVOLUTIONS AND THE GUTENBERG MYTH

1. THE STRUCTURE OF TECHNOLOGICAL REVOLUTIONS

We seem always today to be in the midst of one technological revolution or another. We imagine the future shaped by them. And in our accounts of the past, we point to them as significant forces that have given form and direction to our civilizations. Yet nowhere do we find any substantive discussion as to what exactly a technological revolution is. In fact, we refer to them as though their nature were obvious. We make use of historical examples to characterize the appearance of exciting new technologies and to prescribe ways for dealing with their "impact." And we make these comparisons and prescriptions with such alacrity as to suggest that an explication of what we take the form of a technological revolution to be is simply unnecessary. In these various ways we have, nonetheless, developed an implicit characterization of technological revolutions that consistently results in alarmingly faulty histories and a posture toward current technologies that is frighteningly misleading and ultimately irresponsible. My aim here is to begin to surface this characterization, which I call the Gutenberg Myth, and to suggest that we are in need of an historically and conceptually more accurate and prescriptively more responsible understanding of the structure of technological revolutions.

It is telling that the critical discussion of *scientific* revolutions that has ensued since the publication of Thomas Kuhn's landmark book, *The Structure of Scientific Revolutions* (Kuhn, 1962), has gone on with no parallel discussion of the structure of *technological* revolutions. That we have apparently seen no need for such a discussion betrays, I'm afraid, at least two implicit assumptions about the nature of technology and technological revolutions that can no longer stand and we can ill afford. The first is that the structure of technological revolutions can be taken for granted from what

Joseph C. Pitt (ed.), New Directions in the Philosophy of Technology, 63-83.

historians and others have reported about particular cases. A chief aim of the discussion of the printing revolution below is to show that this is an unfounded and misleading assumption. The second is the age-old notion that technology is at root derivative of science and hence anything philosophically or conceptually important about technology must surely "boil down" to something about science. Hobbes so strongly understood science as the "true mother" of technology that he saw any credit given to "the artificer" as a matter of "the midwife passing . . . for the mother" (Hobbes 1958). Our virtual silence with respect to the structure of technological revolutions at a time when the media, from popular to scholarly, seem to insist that we are up to our ears in them, suggests that we hold a similar view.

This latter point is doubly peculiar in technology studies, since there has been a great deal of important work in recent years that significantly recasts our understanding of the basic relationship between science and technology and resoundingly rejects the model we have up until now so devoutly shared with Hobbes. In Don Ihde's work we find a challenge to the traditional historical and ontological priority of science over technology that suggests a kind of leap-frogging of the two (Ihde 1983). Larry Hickman has suggested quite provocatively that a close look at Dewey can reverse the ontological order entirely, leading us to understand science as simply one type of technology–since science for Dewey is but one form of the technological enterprise of productive inquiry (Hickman 1990). The work on scientific communities by Sharon Traweek (1988) and Bruno Latour (1987) depicts an essential and intimate relationship between the doing of science and the use of instruments. In this latter vein, I would argue that when science and technology are understood as practices, it is often difficult to draw any useful or meaningful distinction between them at all.

Such challenges to the traditional distinction notwithstanding, we still tend to speak of scientific revolutions and technological revolutions differently. We treat them implicitly as though they were distinct. We offer different examples of each. If for no other reason, shouldn't our understanding of the structure of technological revolutions change given our changing understanding of the relationship between science and technology?

I call our implicit model of the structure of technological revolutions the Gutenberg Myth because more than other episodes in the history of technology, the advent of movable type printing is pointed to as an example of what a technological revolution is. Moreover, the printing revolution has become the exemplar (to follow Kuhnian terminology) of technological revolutions in general; it is the example most commonly given as an instance of a technological revolution when characterizing other technologies as revolutionary (this has been seen most recently, for example, in discussions of the personal computer and the "data highway").

The model of technological revolutions implicit in the Gutenberg myth is simple in structure. It depicts rapid, far-reaching social change as issuing from the introduction of a single new technology. The model is one-dimensional, causal and technologically deterministic. It focuses on a single technology that is characterized as the sole cause of the social changes associated with it. In this view, just as technology is understood as proceeding from science, so is dramatic social change seen as proceeding from a new technology. The case of movable type printing has been treated along these lines almost without exception. And as will be seen in the narrative on printing below, although this structure can be outlined simply, the tenacity with which the Gutenberg myth holds onto our imaginations and shapes our understanding of technology and social change are not so easily surfaced and dismissed.

In keeping with the Gutenberg myth, treatments of the printing revolution in the general literature of encyclopedias, textbooks, and journalistic media are generally one-dimensional, factually in error and conceptually misleading. Such accounts commonly hold that the invention of movable type brought about very rapid, far-reaching social change through the spread of literacy and learning to a broad population. However, as I will show, the minimal social and material elements for such a revolution were nowhere to be found at the time that movable type made its appearance – indeed, they only came about over the course of generations. Nor was the eventual rise of mass literacy in Europe solely the product of printing or any other technology.

In the specialist literature (technological studies and philosophy of technology in particular) there is, as I have noted, a lack of solid discussion as to what our understanding of the structure of technological revolutions ought to be or even what subjects should be under debate. References to the printing revolution nonetheless abound.

Given the faultiness of the traditional model, discussions of technological revolutions and analogies between and among them rarely provide the sort of insight that a disciplined understanding of them requires.[1] This has become of growing importance as discussions of new technologies have increasingly included predictions of vast, technologically-created social change and urgent prescriptions by "experts" as to what non-experts must to do to adapt to this deterministic future. In treating the social change associated with new technologies as "caused" by the appearance of technologies, for example, this version of technological determinism discourages discussion of moral or political responsibility with respect to the shape and course of those social changes.

What is needed, as I posit by way of the following narrative, is a model that characterizes technological revolutions not as *caused* by single technological innovations but as *constituted* in multiple, mutually-influencing technological and *social* innovations. Such a model would treat the likes of human choice and shifts in social, moral and political values as significant factors within technological revolutions. Conversely, it would open our present discussions of important new technologies to considerations of social, moral and political responsibility with respect to the choices we make (or fail to make) about their design and use.

The following narrative on the printing revolution, then, has two aims: first, to add a focused interpretation of the printing revolution to those broader efforts noted above that challenge the misleading understanding of technological revolutions implicit in the literature; second, to offer a case in point for advancing an alternative understanding of technological revolutions that is historically and conceptually on firmer ground – in particular, one that supports a more responsible posture toward current and historical instances of technology and social change. If we are to assume such a posture, discussions of technology and change in the general literature must change. And our understanding of the structure of technological revolutions in technology studies and in the philosophy of technology, especially, ought to facilitate that change, at least by challenging the dominance of the Gutenberg myth.

2. THE GUTENBERG MYTH

The appearance of printing has been celebrated as a major step in the advance of technology and the development of Western Civilization. Charles Babbage, the 19th century inventor of calculating machines, remarked that "the modern world commences with the printing press."[2] Today, it is commonly held that the invention of movable type made possible the mass distribution of the printed word, that it brought about a string of broad and rapid social advances through an explosion in literacy and learning, and that ultimately it "democratized knowledge" (Burke 1978). *The McGraw-Hill Encyclopedia of Science and Technology* states that the invention of movable type printing "was one of the most important inventions in human history. It was significant, even revolutionary, in two respects." The first was the principle of movable type itself. "A second and more important aspect [was that it] . . . made it possible to put more information into the hands of more people in less time and at lower cost, and thereby to spread literacy and learning more widely and rapidly than ever before" (Bruno 1987). As historian of science Derek de Solla Price put it, "By 1500 . . . the printed book had become a quite new force. The momentous effect, of course, was that the world of learning, hitherto the domain of a tiny privileged elite, was suddenly made much more accessible to the common man" (de Solla Price 1975, p. 99).

This is the traditional view of the Gutenberg revolution. It is the image that comes to mind when printing is suggested as a model for other technological revolutions. It does not, however, reflect the realities of European history. The process by which the world of learning became "accessible to the common man" entailed several factors in addition to movable type. I will focus on two of them that are particularly central. The first was paper: bringing the printed word to the masses required a medium to print on which was available in quantities and at a cost amenable to "mass distribution." Second, the "common man" needed to be able to read. Neither of these requirements obtained in 1500. These discrepancies point to the need for a richer, more accurate account of the story of movable type and its relationship to the spread of literacy and learning to a broad population in Western Civilization. It is, after all, the achievement of broad literacy and learning that makes the story of printing revolutionary. And this story can begin with the Gutenberg Bible itself.

3. GUTENBERG REVISITED

The invention of movable type printing is usually credited to Johann Gutenberg, who used it to produce one of the most famous masterpieces of the 15th century: the book now called the Gutenberg Bible. As with many important inventions, when, where and by whom the invention actually came about has been the subject of considerable debate. One major theory, for example, holds that the first person in Europe to develop movable type was not Gutenberg but Laurens Janszoon Coster, a contemporary of Gutenberg living in the Netherlands. (This theory has a good measure of historical merit, as well as a great deal of popularity among the Dutch.) In fact, Gutenberg was not the only person in Europe working on the idea of mechanical printing. It is also quite likely that he drew on the work of craftsmen in allied areas such as metal alloying and casting. Further, it is possible that the work of Gutenberg, Coster and others in Europe owes a significant debt to the traditional use of print blocks in Asia, which by the 15th century had been a common practice in China, Japan and Korea for generations – including their use in the printing of books. With such qualifications, the advent of movable type printing in mid-15th-century Europe can be taken as a starting point of the "print revolution," and this is associated, symbolically at least, with the printing of the Gutenberg Bible.

The principle of movable type used in printing the Gutenberg Bible is elegantly simple. Gutenberg took as his basic unit the single letter, making type blocks that carried one letter each. The blocks were of standard size in all their dimensions so they could be set one by one, in any combination, into consistently straight lines on a flat printing bed in an arrangement of text making up an entire page, which in turn offered a consistently level printing surface. After being used to print any number of copies of that page, the arrangement could be broken down and the type blocks reused. Gutenberg's chief technological accomplishments included developing techniques of metal casting and alloying that made this process economically and technically practical.

In keeping with the principle of movable type, each letter of the Gutenberg Bible was set by hand. The pages, over one thousand of them, were printed one at a time onto hide parchment or rag-fiber paper. Then, in the spirit of the manuscripts of the day, hand illuminations and lettering details were added to each of the printed pages. Letters at the beginning of chapters, for example, were written in by hand and elaborately illumined and every capital letter in the entire text had a red highlight scribed through it. Finally, the leaves were bound, again by the hand of the craftsman, into large, elegant volumes. The whole project, from casting the first type to binding the final volume, took several years.

Looking at the Gutenberg Bible and at Gutenberg's practice as a craftsman, one does not sense the invention and exploitation of a revolutionary technology but rather the conservative use of an innovative technique for the mechanical production of manuscripts. In fact, the very aim of movable type printing and the technical innovations Gutenberg brought to it were directed at producing volumes that looked as much as possible like manuscripts. Within the printing craft in general, moreover, the same standards of craftsmanship and aesthetics associated with manuscripts were applied to printed books for at least two generations beyond the Gutenberg Bible. Had the new technology not met these old standards, it might well have been dismissed as a failed experiment. Indeed, a judgment like this seems to have been made in China: type blocks containing a single character had been developed, but were abandoned in favor of plate blocks containing entire texts because the Chinese felt that the type blocks were not sufficiently consistent in shape to produce a printed page that met the high aesthetic standards of their calligraphy. This suggests that the initiation of a technological revolution may often depend on *new* technologies being functionally compatible with *old* craft practices and *traditional* values of the surrounding culture.

Since printing the Gutenberg Bible on fine parchment entailed the use of hundreds of pieces of hide, raw materials alone could be a costly factor. A single copy of the Gutenberg Bible, for example, could require the skins of fifty to seventy-five sheep. Likewise, the paper Gutenberg used (most copies were on paper), though less costly than parchment, was nonetheless a valuable commodity. It was produced a single sheet at a time by skilled craftsmen who were dependent upon often scarce supplies of waste rag or scrap from garment manufacture.

Gutenberg's finished product, then, was a marvelously fine piece of craftsmanship, a beautiful and valuable example of the printer's art. And it remains so to the present day. But it was also an artifact for elites, chiefly the aristocracy and the Church. The Word as printed by Gutenberg was not distributed to the masses. In fact, only about two hundred copies of the Gutenberg Bible were ever produced. Meanwhile, printed books in general remained inaccessibly costly to de Solla Price's "common man" well beyond 1500.

Not only were most people in the 15th century unable to buy a Gutenberg Bible, they were also unable to read. This is an essential element in assessing the course of the printing revolution, since any potential mass social change associated with printing must be weighed against the extent of mass illiteracy.

Though difficult to define and measure, even a cautious estimate places illiteracy in 15th century Europe at well over 90% of the general population.[3] Further, since those who could read were concentrated among clergy, scholars and aristocrats, and there almost entirely among men, illiteracy in most other segments of society was close to universal. Even among those who were able to read, not all were schooled in Latin, which was, of course, the language of the Bible.

In the light of such factors, Gutenberg's efforts appear far from revolutionary – indeed, they are quite in keeping with the social, religious, political, and economic institutions of his day.

4. PRINTING AND MASS ILLITERACY

Nonetheless, printing was much faster and less expensive than the calligraphing of manuscripts. And this, along with the aesthetic quality of printed texts, came to present a significant challenge to traditional manuscript production. Indeed, the displacement of calligraphy by movable type as the preferred means of creating documents resulted in a dramatic spread of printing technology. By the middle of the 16th century, presses were operating in most major centers of Europe and had begun to appear in the Middle East and Asia.[4] In the New World, the first press was established at Mexico City in 1539, the second, a century later in 1638 at Cambridge, Massachusetts.

Within literate communities, the press helped bring about a greater availability of texts and a vastly expanded number of individual titles. It is estimated that some thirty thousand editions were printed in the first fifty years following the advent of the press, equaling the number produced in the previous one thousand years. Moreover, in scholastic communities, the press became a key element in a remarkably rapid growth in the translation and publication of classical texts – a phenomenon which was to fan the glowing coals of the Enlightenment.[5]

By 1650, movable type technology was a fact of life in hundreds of communities throughout Europe. But the spread of the technology did *not* carry with it a broad social revolution in literacy and learning. Although levels of literacy had risen in some sectors of the population (in connection with elements of the Reformation, for example), illiteracy in Europe in general stood at about 80% in 1650, and the ability to read and write remained significantly associated with class and gender. At this point, two hundred years after Gutenberg, access to the world of learning was *not* a fact of life for the masses of people.

Coming into the 18th century, illiteracy began to fall at a greater rate. By 1700, it had fallen to between 65% and 70%. As the 19th century approached, the figure neared 50%. Among the factors associated with this improvement was the growing role of reading and writing in work, especially in the trades. In fact, one of the earliest exceptions to the exclusion of literacy to all but social elites was its appearance among the trades. Reading and writing became a growing part of work as the increasing organization of society expanded the need to keep records. Treatises on craft skills, meanwhile, became one of the largest categories of printed books (coming in third after religion and law).

Significantly, the spread of literacy to more people and more classes of people during this period (approximately 1650 to 1800) occurred alongside some rather remarkable changes in both the idea and reality of social equality. The notion that individuals are equal within a human community found strong expression in the political, literary and philosophical writings of the time. Philosophical propositions like Hobbes's argument that all people are equal within "a state of nature" and Locke's contention that all people are by nature equally free, found broad expression in Jefferson's declaration that "all men are created equal" and Rousseau's insistence that social inequalities are unnatural creations of our institutions. Claims of a political right to specific forms of social equality became banners of the American and French Revolutions. Later, Kant gave these themes a stronger moral sense with his argument that all people are (or ought to be) moral equals as legislators within a "kingdom of ends." Similarly, various notions of equality were explored in the art, architecture and music of the period.

Ultimately, this sweeping reevaluation of the idea of equality entailed nothing less than a recasting of what was understood to be the entitlements of membership in society and the beginnings of a parallel redesign of the social institutions associated with the provision and guaranteeing of those rights. And all this made the spread of literacy, as a perquisite of citizenship rather than of privilege, more "thinkable" and thus more feasible. The decline in illiteracy during this period, therefore, must be understood as much more the product of changing social, political and moral values than a consequence of the continued presence of the centuries-old technology of printing. That is, this aspect of the printing revolution was due more to social than technological factors.

5. THE PAPER IT'S PRINTED ON

From the time of Gutenberg to the beginning of the 19th century, paper was the product of costly materials and highly skilled craftsmanship. No major advance had been made in the painstaking, sheet-by-sheet hand work of the ancient craft since it entered Europe from the Islamic world in the 12th century. [6]

Correspondingly, the value placed on paper is reflected in numerous laws and customs from the time of Gutenberg to the 19th century. A 1666 British law, for example, held that only wool, being unsuitable for papermaking, could be used in burying the dead. An early New England periodical, encouraging readers to save rags for papermakers, suggested that each housewife make a rag bag and keep it next to the family Bible – a wonderfully symbolic connection emphasizing the value placed on paper. The *Boston News Letter* printed an announcement in 1769 that "the bell cart will go through Boston about the end of each month" for the purpose of collecting rags. Readers were further encouraged to fulfill their civic duty with the following poem.

Rags are beauties which concealed lie,
But when in paper, how charms the eye!
Pray save your rags, new beauties to discover,
For of paper, truly, everyone's a lover;
By the pen and press such knowledge is displayed
As wouldn't exist if paper was not made.
Wisdom of things, mysterious, divine,
Illustriously doth on paper shine.

The Massachusetts General Court required in 1776 that in each community a person be appointed to collect rags. To conserve paper, it remained until 1818 a punishable offense in England to print newssheets or broadsides larger than 22 by 32 inches (roughly the current size of the *New York Times*).

The search for a reliable, less expensive alternative to rag scrap as the raw material for paper had been unsuccessful – though not for want of effort. Several materials were tried experimentally, including hemp, pine cones, used shingles, potatoes, and asbestos. A particularly imaginative source for rag itself was found in the mid-19th century in Egyptian mummies. Shiploads of mummies were sent from Egypt to paper companies in the United States, where their linen wrappings were taken off and recycled, so to speak, into paper. This continued for some time, apparently without the intervention of concerned health officials, outraged clergy or jealous archaeologists. The only competition the papermakers had for the mummies was from the new Egyptian railroad, which, it seems, used them as fuel.

That mummies could be shipped all the way from Egypt to the U.S. solely for the use of their wrappings is a gauge of the value and cost of paper at the time (as well as measure of deficit in respect for the dead). The availability of rag and the cost of paper were clearly serious obstacles to printing on a mass scale well into the 19th century.

6. MASS PRINTING AND MASS SOCIETY

Through the course of the 19th century, several technological and social innovations combined to produce dramatic changes in printing and literacy. By the turn of the century, wood pulp had been identified as a reliable and plentiful fiber source for papermaking.

The first experimental techniques for producing pulp-fiber paper by machine were operational by the 1810s. In the 1840s, a machine designed specifically for the mass production of paper-quality pulp was introduced. By the 1860s, wood pulp paper was being made commercially in mills that could turn out paper at dazzling speed in continuous rolls of indefinite length. The development of machine-made pulp and paper made possible a rapid drop in the cost of paper. The price of newsprint, for example, fell by a factor of ten between the 1860s and 1890s.

Printing itself developed remarkably through the 19th century. Steam power was introduced to press operation in 1810. In the middle of the century, the flat reciprocating printing bed was displaced by the development of the stereotype cylinder, the molded and curved printing surfaces now associated with the "rolling press." A major step in the printing craft was made in 1884 with the invention of the linotype. With this machine, the highly time-consuming hand work of setting type one letter at a time could be replaced by a mechanized operation that cast type by the line. This principle dramatically increased the speed and reduced the cost of type setting. In doing so, the linotype became an important factor in making possible the mass production of the printed word. With respect to printing technology, meanwhile, the linotype represents a significant step away from the principle of movable type, since it treats the line rather than the single letter as its basic unit.

Resting on expanding notions of social equality, a fundamentally important social innovation of the 19th century was the introduction of public education. From the middle of the century onward, state-supported elementary education became a reality in country after country. By the end of the century, it was accepted, at least in principle, throughout Western Europe.[7] The parallel decline in illiteracy was dramatic. In 1800, half of the general population was illiterate. By the century's end, the figure was under 10%.

With the advent of mass literacy and the development of technologies for the rapid and relatively inexpensive mass production of paper and printed materials, the mass distribution of the printed word became a meaningful and practical possibility. It was not until this point in the 19th century that the social and technological elements for making "the world of learning . . . accessible to the common" person through printing were fully in place – four hundred years after the appearance of the movable type press.

All of these factors combined begin to sketch out a fuller, more accurate history of the understanding of the printing revolution; they also suggest the need for a broader, multi-dimensional model of the structure of technological revolutions in place of the Gutenberg myth. At the very least, this account of the printing revolution suggests that the traditional, one-dimensional model of new technologies (or a single new material gadget) causing broad social change must be regarded with deep suspicion.

A healthy suspicion about the Gutenberg myth is also called for in our efforts to understand the social changes associated with new technologies, both historical and current. At the very least, we need a new model of the structure of technological revolutions in order to assess and understand the role of technological change in our lives and cultures in a way that avoids the political and moral myopia that the Gutenberg myth encourages. It is to the exploration of some examples of these pitfalls that I will now turn.

7. ILLITERACY IN THE SHADOW OF TECHNOLOGY

The story of the spread of literacy and learning in Western Civilization is an important one, as is the role of the printing press within it. At the same time, the call for making "learning . . . accessible to the common" person is still heard today in the broader context of making literacy and learning accessible to the peoples of the world.

Approximately 90% of the illiterate people alive today live in the developing world, where the average level of illiteracy is around 40% of the adult population. Many of these countries have literacy programs, and there are internationally sponsored literacy initiatives through the United Nations and private organizations. Progress is being made, but neither the "problem" nor the "solution" is simple. Nor is it simply a matter of technological need. The level of illiteracy, for example, fell by almost 10% between 1970 and 1980. Given population trends, however, the absolute number of illiterate people *increased* during that period by nearly seventy million.

Just as the Gutenberg myth fails to capture the realities of Western history, so it fails as a model for assessing the current realities of the developing world; the image of literacy and learning as emerging on a mass scale from the introduction of printing technology would, indeed, be a very misleading model for developing nations. The situation in the developing world is neither well understood nor effectively addressed by assuming that mass literacy can emerge from the introduction of a single technology. Today, print and other technologies such as computers are readily available, but this has not made it possible to "put more information into the hands of more people . . . and thereby to spread literacy and learning" throughout the developing world (as the McGraw-Hill encyclopedia's formulation of the Gutenberg myth might suggest).

In addressing illiteracy in the developing countries, the role of printing technologies can be essential – as can be the need for the associated maintenance skills, spare parts and trained labor. But the major factors identified by these countries as obstacles to the achievement of broader literacy are *not* technological. At the center, for example, is the issue of language. In many developing countries the difficulties in achieving general literacy are multiplied by the existence of several language groups. Some countries have no language policy; some recognize both a national language and local languages or dialects. Often matters of language rest on delicate ethnic or political factors. In some language groups there is need for professional writers. A general need is expressed for primers and curricula which are locally appropriate in terms of social, religious or moral values and with respect to concerns for fostering development aims. There is a continuing need for reading materials for the newly literate. Many countries identify a shortage of trained teachers fluent in target languages, as well as ministerial personnel to plan and administer literacy programs. Underlying all of these is the ongoing problem of limited funds (UNICEF-UNESCO Joint Working Group 1984).

Illiteracy is a serious concern for many developing countries. At the same time, there is a growing awareness of the loss or disruption of cultural traditions that the institutionalization of mass literacy can entail. Indeed, the spread of literacy (in the developed as well as the developing world) has often displaced or threatened such cultural elements as non-written languages, forms of know-how transmitted through the practice of crafts, and modes of social and political discourse carried out through public recitations in traditional oral forms. Needless to say, those who control the tools of such processes hold, even if unwittingly, a most elemental form of political power over those who don't.

Such issues would be significant matters in any culture. They entail complex, mutually-influencing social and technological realities, which an unexamined, gung-ho endorsement of the deterministic and one-dimensional perspective of the Gutenberg myth encourages us to ignore.

8. LITERACY WITHOUT PRINTING

It is clear that movable type technology was a key factor in the general spread of literacy in Europe. But even in Europe, movable type was not the only formula for bringing literacy to a broad population. Indeed, a respectable level of literacy was achieved in several instances without the use of a press.

For centuries *before* the advent of the press, in Islamic Spain and in Jewish communities throughout Europe, for example, the ability to pass a reading test was a prerequisite for entering manhood. Moslem and Jewish boys were expected to learn to read holy books, and their communities were expected to provide teachers and institutions to make this possible. And indeed they did, and to a remarkable degree. Moreover, in Islamic communities, in a general sense, the language of the mosque was the language of the market, a social advantage to literacy in Arabic. In Jewish communities, meanwhile, though women were generally excluded from studying holy texts, it was not unusual for women to acquire at least a rudimentary reading ability in the vernacular, along with basic ciphering skills, by virtue of their involvement in commerce. Women and girls worked in business, in part to provide a measure of free time for men and boys to study. In these communities, not only was the level of literacy on an order several times greater than in Christian Europe, but the ability to read and write was not the sole province of elites (with the general but not absolute exception of gender).

This means that well before the printing press, there existed in Europe two models of socio-technical organization, two alternative technologies, that sustained literacy at a level which was not to be reached in the general population until roughly 400 years after the appearance of movable type. Yet the model of technological revolutions embodied in the Gutenberg myth, with its lopsided sense of the relationship between technology and social practice, blinds us to this history. Accordingly, the McGraw-Hill encyclopedia's claim that printing technology "spread literacy and learning more widely and rapidly than ever before" betrays, at the very least, a need of a culturally more modest sense of European history. In general, the technological tunnel vision of the Gutenberg myth leaves us in need of a broader understanding of the varied forms that the relationship between technology and social practice can take.

9. EXCITING NEW TECHNOLOGIES, SAME OLD MYTH

The printing revolution is often evoked as a model or exemplar for understanding the social importance of new technologies. The structure of technological revolutions implicit in such analogies is typically that of the Gutenberg myth: a single technology is taken as the sole cause of rapid and far-reaching social change. Throughout the 1980s, for example, the personal computer was depicted in the popular and scholarly press as singlehandedly revolutionizing the whole of society. By the early 1990s, the same was being said about the "data superhighway." Not surprisingly, such discussions of exciting new technologies reflect the same sorts of historical and conceptual distortions found in nearly all references to the printing revolution.

Throughout the 1980s, the computer revolutionaries told us that computers would give us access to the world of information and that computer literacy had, therefore, become an essential skill "[j]ust as printing made the written word accessible to all and the ability to read and write a necessity and not a privilege" (Galanter 1984, p.14). However, as even this modest look beyond the Gutenberg myth indicates, the spread of printing throughout Europe did not make the written word accessible to all, nor did it alone make literacy necessary or even possible. Change a few key words and suddenly it becomes access to the data highway that is the key to the information promised land. That proponents of such claims have so often sustained them through appeals to the Gutenberg myth does little for their credibility. Yet the parallel is rarely challenged. And the causal, deterministic, one-dimensional model of the structure of technological revolutions goes on.

In the case of the personal computer, the model of broad social change issuing from a single technology encouraged wildly exaggerated and misleading claims about computers "revolutionizing practically every facet of our society" (Bitter and Camuse 1984, p.

22). And once this technologically-driven revolution was declared, prescriptions for what it then required of us quickly followed. It was commonly argued throughout the 1980s, for example, that the spread of computers in our society had made it "imperative that schools implement computer literacy curricula" (Bitter and Camuse 1984, p. 262). Indeed, it was claimed that computer literacy had become "as basic as the need to read, write and solve mathematical problems" (Bitter and Camuse 1984, p. 262), and that "to be culturally literate in our society, one must also be a computer literate" (*PC Week* 1987).

Such claims, however, have little basis even in the nature of computers themselves. There is nothing about the spread of computers in our society that has made it necessary for everyone to be trained in how to use them. In fact, one factor cited by the computer industry as key in the successful spread of computers has been their use requiring less and less specialized know-how; computers have become increasingly common as their design has allowed more and more "computer illiterates" to use them. In fact, by the late-80s, many educators, parents and employers had begun to wonder (and for good reasons) whether they had been oversold on computers and computer literacy.

Nonetheless, frenzied predictions and prescriptions about a computer revolution have flourished. Such claims typically have been very broad and very urgent. They have tended to be tightly focused on the technology and sweepingly wide in what they take to be the social changes issuing from it. In a telling sense, this treatment of computers, like that of so many other new and exciting technologies, says more about the model of technological revolution they assume than about the specific technologies they address. Such claims, for example, as "every single purpose of our public schools . . . can be more effectively achieved through the intelligent utilization of [this technology]," and that "it deserves a place alongside the textbook as the most important educational advance of the century," would pass without a blink of the eye in a recent article about computers; they were made, however, in the 1950s and 1960s about the classroom use of television.[8] Blink again, and you find them being made in the 1990s about the data highway.

We have, in fact, applied the Gutenberg myth to one new technology after another. If nothing else, a look at the structure of the printing revolution suggests that we need to be more cautious in assessing our new technologies and in making predictions and prescriptions about what they require. When the computer revolutionaries proclaimed that we all *must* become computer literate in order to be members of an emerging information society, we should not have heard the call to join in an exciting new technological revolution, we should have heard a tired old refrain. In the last analysis, neither the computer nor the data highway alone has the power to give or withhold membership in an information society. Experts who claim that they do are not pointing out the future for the rest of us so much as they are making a political power grab, whether intended or not, for control over what that future will be. When the next new and exciting technology comes along and the call to arms goes out, perhaps we will pass up the rush to do the inevitable and instead stop and ask ourselves what is important to us that this new technology might threaten or advance.

Just as the Gutenberg myth fails to capture the realities of the printing revolution, so did it fail to temper predictions about the revolution television was to bring to the classroom. I doubt that it will have proved to serve us well in evaluating and fashioning a role for computers or the data highway in our society.[9]

10. CONCLUSION: REVOLUTIONS RECAPTURED

This brief look at the printing revolution suggests that the structure of technological revolutions is neither simple, technologically determined, nor everywhere the same, either historically or culturally. It is clear that the model of sweeping social change as caused by a single technological innovation is historically and conceptually faulty and misleading. Such changes are not *caused* by the appearance of a single gadget; they are *constituted* in multiple, mutually-influencing technological and social innovations. A new model of the structure of technological revolutions must reflect these facts. Moreover, it must be able to embrace the deeply held values that inform our cultures and underlie the choices we make about the direction our technologies ought to take – even when those choices are made by default rather than by design. The technological determinism of the Gutenberg myth forecloses discussion of just these social, political and moral values, which are ultimately the only means by which we can distinguish between an appropriate and inappropriate role for any technology.

If we continue to base our understanding of the structure of technological revolutions on the Gutenberg myth, we do so at the risk of undermining our ability effectively and responsibly to address the complex social and technological realities which the many technological revolutions before us are certain to entail.

San Jose State University

NOTES

1 An exception to the traditional treatments of technological revolutions can be found, for example, in Basalla 1988. However, Basalla does not go far enough, in my view, in linking social and technological factors (see, for example, his treatment of progress in the final chapter). For an initial exposition of these themes, see Cook 1985.

2 The quotation appears without citation in Birdsall and Cipolla 1977, p. 111.

3 The difficulties in defining literacy and establishing levels of illiteracy are well covered, for example, in Cressy 1980, and Pattison 1982.

4 A very detailed account of the early years of the printed book can be found in Febvre and Martin 1976.

5 A landmark study of the role of printing in *literate* communities is Eisenstein 1984.

6 For a thorough and delightful study of the history and craft of papermaking (from which many of the examples given here are drawn) see Hunter 1957.

7 For a discussion of the rise of mass society, with reference to the role of state supported education, see Biddis 1977.

8 Classroom Television was a frequent topic in popular and scholarly publications in the late 1950s and early 1960s. The quotes given here are, respectively, by Wilson Cahall, director of the famous Hagerstown teaching experiment, quoted in Weiss 1957, and William M. Brish, Washington County (MD) School Superintendent, quoted in Bakal 1957.

9 For an exploration of political implications of the common treatment of the computer revolution, see Winner 1984. For a more general critique see Cook 1990.

REFERENCES

Bakal, Carl (1957). "The Schools of Tomorrow." *Saturday Review* 24 August: 9-11 and 35-37.

Basalla, G. (1988). *The Evolution of Technology.* Cambridge: Cambridge University Press.

Biddis, Michael D. (1977). *The Age of the Masses.* New York: Harper and Row.

Birdsall, Derek, and Carlo M. Cipolla (1977). *The Technology of Man.* UK: Penshurst Press.

Bitter, Gary G., and Ruth A. Camuse (1984). *Using a Microcomputer in the Classroom.* Reston, VA; Reston.

Bruno, M. (1987). "Printing," in *The McGraw-Hill Encyclopedia of Science and Technology,* volume 14.

Burke, James (1978). *Connections.* Boston: Little, Brown and Company.

Cook, Scott D. N. (1985). "Technology and Social Change: The Myth of Movable Type," in *The Weaver.* 4:1.

Cook, Scott D. N. (1990). "Computers and Rumors of Revolution." *The Journal of Computing and Society* 1: 29-41.

Cressy, David (1980). *Literacy and the Social Order.* Cambridge: Cambridge University Press.

de Solla Price, Derek J. (1975). *Science Since Babylon.* New Haven: Yale University Press.

Eisenstein, Elizabeth (1984). *The Printing Revolution in Early Modern Europe.* Cambridge: Cambridge University Press.

Febvre, Lucien, and Henri-Jean Martin (1976). *The Coming of the Book.* London: NLB.

Galanter, Eugene (1983). *Kids and Computers.* New York: Perigee Books, The Putnam Publishing Group.

Hickman, Larry (1990). *John Dewey's Pragmatic Technology.* Bloomington and Indianapolis: Indiana University Press.

Hobbes, Thomas (1958 reprint). *Leviathan.* Indianapolis: Bobbs- Merrill.

Hunter, Dard (1957). *Papermaking: The History and Technique of an Ancient Craft.* New York: Knopf.

Ihde, Don (1983). *Existential Technics.* Albany: State University of New York Press.

Kuhn, Thomas S. ([1962] 1970). *The Structure of Scientific Revolutions.* Chicago: University of Chicago Press.

Latour, Bruno (1987). *Science in Action.* Cambridge: Harvard University Press.

Pattison, Robert (1982). *On Literacy: The Politics of the Word from Homer to the Age of Rock.* Oxford: Oxford University Press.

PC Week (1987). Editorial. "Funding Computer Literacy Pays." August 25.

Traweek, Sharon (1988). *Beamtimes and Lifetimes*. Cambridge: Harvard University Press.

UNICEF-UNESCO Joint Working Group (1984). *Moving Toward Universal Primary Education and Literacy*. UNICEF Occasional Papers Series no. 4. New York: UNICEF.

Weiss, John K. (1957). "Educational TV: Teacher's Friend." *Saturday Review*, 14 September.

Winner, Langdon (1986). "Mythinformation," in *The Whale and the Reactor*. Chicago: University of Chicago Press.

DANIEL DeNICOLA

PARADIGMS AND PARAPHERNALIA: ON THE RELATION BETWEEN THEORY AND TECHNOLOGY IN SCIENCE

Not only are good experiments based on theory, but even the means to perform them are theory incarnate.

– A. Koyré

What is the connection between theory and technology in science? What is the relationship between the various activities of "doing" science and the instruments that enable these activities? My interest here is to explore these questions in a very broad and elementary way, occasionally citing examples plucked from the history of science.

Perhaps it seems that I am in pursuit of the obvious. Who would doubt that there is a strong tie between scientific theory and technology? The relationship is reciprocal: scientific discoveries often yield technological advances, and new technology permits further discoveries. Although this account is accurate, it is superficial. My analysis of the relationship will lead me to advance a strong claim. I believe that theory and technology are so intertwined in science that neither can be studied independently of the other without distortion and diminishment of understanding. Ultimately, I want to argue that a deeper understanding of this relationship is essential to an adequate philosophy of science and to science education, although I will not address that topic here.

1. THE TECHNOLOGY OF DISCOVERY

Science is an enterprise that requires paraphernalia. The image of the scientist presented in the movies and other media of popular culture is that of the researcher in a laboratory or in the field surrounded by scientific paraphernalia. Faust in his study. Dr. Frankenstein in his laboratory. The heroic scientists of so many science fiction films. It is the apparatus and materials that surround them – the microscopes, retorts, Bunsen burners, oscilloscopes,

Joseph C. Pitt (ed.), New Directions in the Philosophy of Technology, 85-93.

Geiger counters, electromagnets (often arranged in amusingly nonfunctional arrays) – all the stuff that says "science is being done here."

Of course the instruments have changed with the ages – the era of simple weights and measures; the era of optical telescopes and pendulum clocks; the era of Leyden jars and lightning rods; and now the era of radiotelescopes and gigantic scintillation counters. The pursuit of science requires more sophisticated and more expensive equipment with each passing year. Sadly, the days of "table top science" seem gone forever.

To begin to analyze the role of these devices, one must begin with some basic distinctions because not all the paraphernalia one is likely to find in a research center have the same relationship to the pursuit of science. Taking inventory of the contents of a laboratory, one might find *scientific objects* – the specimens, substances, and artifacts selected for study by the scientists. One might also find *supplies* – computer disks, solvents, photographic plates, food for laboratory animals. There may be *equipment* that is only indirectly related to the ongoing research – for example, animal cages, high intensity lamps, exhaust fans, protective glasses, and so on. Finally, one may locate the devices that are of central interest here, the *scientific instruments* – the galvanometers, spectrophotometers, cloud chambers, mass spectrometers, and so on (some of them much too large to be portable – for example, electron microscopes, superconducting supercolliders, and radiotelescopes).[1]

Such scientific instruments may function in several ways. They extend our senses. They detect the occurrence of a phenomenon. They weigh, count, measure, or otherwise assess some dimension of a phenomenon. They may record these observations. They may use data to construct a representation or image of a phenomenon. Some simple instruments do only one of these things; other instruments combine operations. Some instruments alter one or more properties of a phenomenon in predetermined and measurable ways. In all cases, they provide purposefully mediated contact between the researcher and Nature.

These devices represent ingenious combinations of our knowledge of the world and our "know-how." The development of high-powered optical telescopes, for example, requires not only a knowledge of optics but also sophisticated processes of lens grinding and the preparation of special materials. This is the conventional picture: scientific knowledge grows; it is applied by ingenious inventors often aided by serendipity and engineers to create new technology; this technology is then used in scientific research to extend our knowledge. In research, instruments are used to confirm or refute a theory or hypothesis. The introduction of instruments into research reflects the empiricism of science, as they transform the unobservable into the sensory data of meters, graphs, indicators, gauges, etc. Instruments seem neutral and impersonal. They provide observations devoid of researcher prejudice. The data provided are objectified: "It's right there on the film!" "We've recorded a significant fluctuation; look here at the meter." "No doubt you've gained material: look at the scales." Often scientific instruments yield results which violate researcher expectations or carry the experimenter far beyond current knowledge or purpose. One thinks of Galileo, Leeuwenhoek, and Röntgen as observers who were astonished by the phenomena revealed by their instruments. Occasionally, however, instruments may advance our knowledge by their *failure* to detect a phenomenon. One thinks of Michelson and Morley failing to detect the ether wind.

On this simplistic view, then, scientific instruments seem logically independent of scientific theory and the commitments of the experimenter – indeed, their effectiveness in experiments seems in large part to rest on that independence. The levels of theory and observation are thought to be quite distinct. The relationship appears only a matter of the practical relevance to the phenomena under study. The theory, the technology, and the observer are each in its own place and isolable from the others.

However, as soon as this account is given, potentially significant qualifications and reservations must be admitted. First, for an instrument to function properly, it must be *in good working order*. Determining whether a given instrument is in good working order requires a diagnostic routine that is independent of the instrument's immediate application. For example, suppose a technician applies a Geiger counter to a substance and receives a reading of intense radioactivity. The technician may doubt whether the Geiger counter is in good working order. To test it, the technician may apply it to other substances of known levels of radioactivity for comparison – levels known to the technician – or some other independent diagnostic test will be applied. Secondly, instruments must be properly calibrated. Calibration is a complex technical problem that requires the application of independent standards, i.e., a comparison with known and certified values. In practice, portable standards are applied for comparison to permit instrument adjustment.

Third, instruments function properly only within a range of application, and, fourth, they assume that certain standard conditions obtain. A scale may, for example, give accurate readings for weights of a certain range only – say, from one gram to ten pounds. Objects with less weight or more weight will not be read accurately, if at all, and a very heavy object may damage the instrument. In addition, this scale may assume the gravitational pull of Earth at sea level. Fifth, instruments come with a set of instructions; that is, they have established protocols for their proper use. A device can be in proper working order, accurately calibrated, applied within the proper range of phenomena under standard conditions, and still be dysfunctional if the technician is using the machine improperly. Sixth, many instruments require observer competence. This means that skill and experience are required to interpret the data provided by the instrument. Reading a sonogram or an electroencephalogram, for example, requires interpretive skill. Only some observers are competent observers, and the certification of this competence is a knotty epistemological problem.

Undeniably, the instruments of science do provide a kind of objectivity that occasionally yields findings surprising to the researcher. But the six factors outlined above reveal ways in which the researcher's knowledge, commitments, and skills are implicated in the instruments and their proper functioning. When we design our technology of discovery we reveal, in a way, what we expect to find with it.

2. INSTRUMENTS AND HUMAN PURPOSE

"Paraphernalia" is a broad concept. It includes a wide array of devices, including tools, utensils, implements, equipment, machines, appliances, engines, apparatuses, and instruments. And all of these are imbued with human purpose.

In such works as *Personal Knowledge* and *The Study of Man*, Michael Polanyi discussed the nature of machines. He pointed out that "a knowledge of physics and chemistry would in itself not enable us to recognize a machine" (Polanyi [1958] 1964, p. 330). We could produce a complete physico-chemical analysis of the object and still have no sense of whether it is a machine or, let's say, a sculpture. If we were told it was a machine, we might still have no surmise as to its use, and we would certainly have no way to derive its operational status and use from our knowledge of physics and chemistry. We would not know what it was for and whether it was operational. How could we make this determination? To quote Polanyi, "*Only by testing the object practically as a possible instance of known, or conceivable, machines*" can we learn whether the thing serves any purpose and, if so, what purpose (Polanyi, p. 330). That is a fairly ponderous, open-ended test, though it is probably an accurate description of what we would do if, for example, when browsing in an antique shop, we came upon an implement we could not identify. Or what space explorers might do upon finding a suspiciously artificial alien object. "What is this device? What could it be used for?"

In these situations, we cannot make a definitive determination because we lack an important piece of knowledge: knowledge of intended purpose. That insight becomes fundamental in Polanyi's account; to understand a machine is to enter the realm of purpose. Let me quote him at length:

> For a machine is a machine only for someone who relies on it (actually or hypothetically) for some purpose, that he believes to be attainable by what he considers to be the proper functioning of the machine: it is the instrument of the person who relies on it. . . .
>
> Since the control exercised over the machine by the user's mind is – like all interpretations of a system of strict rules – necessarily unspecifiable, the machine can be said to function intelligently only by aid of unspecifiable personal coefficients supplied by the user's mind (Polanyi, p. 262).

According to Polanyi, a machine (for example, a typewriter, an engine, a telephone, or a camera) is characterized by an operational principle. Such a principle is formulable for a patent, which specifies just how the "characteristic parts" work together to achieve the machine's purpose. Such an operational principle extends the conception of the machine to include hypothetical objects of varying sizes, made of a wide range of materials. But the range of things which might be the instances of such an operational principle, the individual objects which might be machines of the same kind, is not specifiable by pure science.

The operational principles define a machine "in good working order." This means that they explain the proper and efficient functioning of the machine. They cannot, however, explain why a machine is "out of order." For an explanation of a particular equipment failure we must turn to science – to physics and chemistry. "The operational principles of machines are therefore *rules of rightness*, which account only for the successful working of machines but leave their failures entirely unexplained" (Polanyi, p. 329).

Like a machine, an instrument is a device designed to achieve an end. Instruments, like machines, have operational principles. However, the term "instrument" suggests a device of delicate construction, or of a precision or capability which extends beyond normal mechanical or manual operation. These connotations hint at the close ties between instrument design and scientific theory.

Many times instruments have been designed for specific research needs. In *The Structure of Scientific Revolutions*, Thomas S. Kuhn lists several examples:

> Special telescopes to demonstrate the Copernican prediction of annual parallax; Atwood's machine, first invented almost a century after the *Principia*, to give the first unequivocal demonstration of Newton's second law; Foucault's apparatus to show that the speed of light is greater in air than in water; or the gigantic scintillation counter designed to demonstrate the existence of the neutrino – these pieces of special apparatus and many others like them illustrate the immense effort and ingenuity that have been required to bring nature and theory into closer and closer agreement (Kuhn [1962] 1970, p. 26).

In some cases, scientific theory has predicted a result years before an instrument could be designed to provide empirical confirmation. Once available and useful in experiments, a piece of technology can in fact come to shape the research agenda.

3. PARADIGMS AND PARAPHERNALIA

The brilliant and influential account of science outlined by Thomas Kuhn in *The Structure of Scientific Revolutions* provides deeper insights into the relation between scientific knowledge and scientific instruments. His central concept – the paradigm – makes impossible a clean separation between theory and observation. "The instance of the paradigm sets the problem to be solved; often the paradigm theory is implicated directly in the design of apparatus able to solve the problem" (Kuhn, p. 27). Kuhn notes that Newton's *Principia* established a paradigm expressed in laws; however, Newton's second law was at the time unconfirmed by empirical observation. Nearly a century later, Atwood invented his machine to give "the first unequivocal demonstration" of Newton's second law. "Without the *Principia* . . . measurements made with the Atwood machine would have meant nothing at all" (Kuhn, p. 27). Instruments sometimes detect paradigmatic anomalies, and they may therefore be essential to the conceptual revolution that replaces one paradigm with another. Röntgen's cathode ray tube produced radiation so unanticipated and unknown it was labeled "X-rays," and it launched a new paradigm.

Some instruments survive the revolution – although their data may be reinterpreted or their ranges of application may be redefined. The revolution from Newtonian to Einsteinian physics, for example, left many instruments of physics intact. Occasionally, the instrument itself is completely revisioned. Consider this example from Kuhn:

> After the assimilation of Franklin's paradigm, the electrician looking at a Leyden jar saw something different from what he had seen before. The device had become a condenser, for which neither the jar shape nor glass was required. Instead, the two conducting coatings – one of which had been no part of the original device – emerged to prominence (Kuhn, p. 117).

Occasionally, some instruments do not survive the revolution and are cast aside.

Instruments are, then, theory-laden artifacts, inexplicable without reference to the theory embedded in their design and operational elements. Consider the Wilson cloud chamber: what is necessary to understand this apparatus? Polanyi would note that, until we understand the purpose for which it is designed, we cannot tell whether it is a machine and, if so, whether it is functioning properly. In this case, however, the purpose of the machine cannot be stated without reference to the content of scientific theory. As is known, a Wilson cloud chamber produces a very moist cloud prevented from condensation by the absence of dust. The cloud reveals the presence of energetic particles passing through the chamber when tiny droplets of water form around the ions produced along the route the particle travels. When the cloud is placed within a magnetic field, information about the mass and charge of the particle may be revealed by its path. The conception of this device rests on the putative existence of unobservable entities postulated in sub-atomic theory. Of course, no observer can see the particle – only the tracks left in the cloud. Such an apparatus is theory incarnate.

4. CONCLUSION

I have provided nothing so splendid as a linear argument. But this cluster of comments, observations, and examples does have a cumulative effect. While scientific instruments are sufficiently independent to present anomalies and refute dearly held hypotheses, they are also linked to theory. First, there are the profound ties of any machine to human purpose. Second, there are the six elements in the effective use of an instrument: (1) good working order; (2) accurate calibration; (3) range of application; (4) standard conditions; (5) protocols of use; and (6) a competent observer. These all display linkages between our knowledge and commitments and the operation of the instrument. Third, there is the deliberate design of instruments to meet a research agenda, incorporating the application of theoretical content. Finally, there is the image of instruments as theory-laden artifacts of our paradigms.

If the relationship between theory and technology in science is accurately described here, it deserves a more central place in the philosophy of science and in science education. There is significant work to be done clarifying the variety of relationships that begin from the vision of an instrument as an embodied theory. Seeing our paradigms in our paraphernalia, and vice versa, may give us an enriched understanding of both.

Rollins College

NOTE

[1] Some instruments, such as a laser, may be ancillary equipment in one instance of research (when, e.g., used in surgery) and a central scientific

instrument in another project (in which the effects or applicability of lasers are researched, for example)

REFERENCES

Kuhn, T. S. ([1962] 1970), *The Structure of Scientific Revolutions.* Second edition, enlarged. Chicago: University of Chicago Press.

Polanyi, M. ([1958]1964), *Personal Knowledge: Toward a Post-Critical Philosophy.* New York: Harper & Row.

ROBERT HAHN

TECHNOLOGY AND ANAXIMANDER'S COSMICAL IMAGINATION:
A Case-Study for the Influence of Monumental Architecture on the Origins of Western Philosophy/Science

I. ANAXIMANDER AND THE ORIGINS OF PHILOSOPHY

How shall we account for the origins of western philosophy, a tradition, following Aristotle, commonly traced back to the Milesian *phusiologoi*, or "nature-accounters," like Thales and Anaximander? The field has been ripe with hypotheses. Aristotle offered a strictly economic hypothesis when he claimed that it was the availability of *leisure*, made possible by the wealth of economic surplus, that proved sufficient to account for the development of speculative thinking.[1] Historians of science, like Sarton, advanced what might be called the *intermingling of beliefs* hypothesis, that the sharing of different ideologies in the interactions with different peoples on the west coast of Asia Minor led to a toleration for differing points of view and consequently an openness in thought for one's own traditional beliefs.[2] The *literacy* hypothesis, advanced by Goody and Watt, claimed that written records provided a sufficient condition to account for the distinct kind of critical evaluation characteristic of Greek philosophy.[3] And others, like Farrington, argued on behalf of *technology*, that technological mastery is a sufficient condition for the development of the critical inquiry that characterizes the emergence of early Greek philosophy.[4]

G. E. R. Lloyd in an important study made the point that although all of these hypotheses may have played a contributing role, none of them was sufficient to account for the origins of Greek philosophy.[5] For while leisure, the intermingling of beliefs, literacy, and technological advance may all have been important, even necessary, to the emergence of western philosophy, had any of them been sufficient we should have expected to see a comparable philosophical tradition emerge earlier in near-eastern civilizations

Joseph C. Pitt (ed.), New Directions in the Philosophy of Technology, 95-138.

such as those of the Egyptians, Babylonians, and other Mesopotamian peoples. For in these cases, all possessed a leisured class, thrived in cosmopolitan cities where the interchange of beliefs must have been great, had literacy (albeit in scribal form connected not to public literacy but to the record-keeping of the central palace), and far surpassed the achievement of the Greek technologies, but not one of them apparently developed a tradition of the self-reflective and critical self-consciousness that is central to the emergence of philosophy and science for the Greeks. Instead, Lloyd joins the chorus of those like Gernet,[6] Vernant,[7] Vidal-Naquet,[8] Detienne,[9] and others, that "Greek rationality in general is the product of the city-state."[10] The *polis* hypothesis is that the radical revision of the framework of political relations and of beliefs about natural phenomena and the world emerged correlatively; developments in the legal and political domain provided images and analogies by which the spheres of law and justice could provide important models for thinking about cosmic order. Thus, the development of the *polis* not only promoted freedom and free speech but also fostered an environment that made possible "radical innovation," the openness of "access to the forum of debate," extolled the "habit of scrutiny," and created an intellectual climate for "the expectation of justification" – of giving an account – and the premium set on rational methods for doing so.[11]

The study presented here concurs with Lloyd's overall assessment, and its modest offering is to draw attention to a slice of this well-formed pie of hypotheses that has been overlooked.[12] This slice is wedged in between the *polis* and *technological* hypotheses. Its thesis is that among the various influences that stimulated Anaximander's philosophical conceptions about the cosmos, the contributing role that the architects and their monumental productions played has been neglected. That Anaximander came to imagine the earth as a column-drum, that the cosmos is measured in column-drum proportions, that monumental stone temple-building was a new innovation in his very backyard at just the time he flourished and "published," and that column-*drum* construction was as new as was its specific technologies, is our starting point. *If* the hypothesis that the architects contributed to Anaximander's philosophical mentality becomes plausible, the next step would be to ask the incisive and ground-breaking questions about the complex of social and political relations in the burgeoning *polis* that motivated the invention of the Greek temple, the building on a monumental scale, and so bring the architect and his innovative technologies onto center stage so that an Anaximander could be significantly influenced. Why build temples? Who paid for them, anyway? And what did the aristocratic patrons believe they were getting for their exorbitant expenditure? It is in the context of the struggle for power in archaic Greece that new and rich insights about the origins of the western philosophical and scientific mentality are yet to be explored. However, in this paper, the case will focus on the plausibility of the contributing role of the architects and their technologies; in another paper, the socio-political details will be more fully explored.

A study of the philosophical mentality of Anaximander of Miletus (c. 610 - 546 B.C.) is by its very nature a study in the origins of western rationality; this short study is part of a much wider project that invites a review of that tradition. The origins of Greek philosophy/science, traceable to the Milesians on Aristotle's account, form the foundation of a tradition that identifies the exercise of human reason as the highest virtue. Familiar studies that have focused on Plato's theory of forms, Cartesian certainty, Kant's *a priori*, and so on, have envisaged western rationality as the triumph of reason over the senses; the mind, not the body, holds the key to a deep understanding about nature. Indeed, it is by transcending the senses and the bodily dimensions of experience that reason can grasp what truly is, *Being* as opposed to *Becoming*. Reason's reflection on its own operations was supposed to be sufficient to generate an understanding of the way things are. A consequence of this position was to define self-consciously the boundaries of appropriate investigations into rationality; since the success of rationality depended upon the conscious rejection of the bodily and sensorial aspects of experience, a knowledge of the historical, social, political, religious, economic, and technological *contexts* was routinely excluded from further examination. The broad thesis I am pursuing is that traditional attempts to understand reason and rationality trans-temporally and hence trans-contextually, disengaged from the body and its situatedness in an historico-socio-political order, have proved inadequate to account for the nature and origins of western philosophy/science. And thus, philosophy cannot understand itself, its purposes and tasks, independent of that embeddedness and our reflection upon it.

Why is this review and reassessment necessary? The recent and important work of Jonathan Barnes, in his two volume study of the Presocratics, would make it impossible to undertake the kind of investigation pursued here. Barnes represents a traditional view endorsed by Descartes, Kant, and others that denies that an understanding of rationality depends in any way upon the historical context in which reason operates.

I have little concern with history. It is a platitude that a thinker can be understood only against a historical background; but that, like all platitudes is at best a half-truth, and I do not believe that a detailed knowledge of Greek history greatly enhances our comprehension of Greek philosophy. *Philosophy lives a supracelestial life, beyond the confines of space and time*; and if philosophers are, perforce, small spatio-temporal creatures, a minute attention to their small spatio-temporal concerns will more often obfuscate than illumine their philosophies. History, however, is intrinsically entertaining. A few external facts and figures may serve to relieve the reader from the purely abstract narrative: I hope that my occasional historical paragraphs may be of use to that end, and may do something to placate the historically-minded reader.[13]

Barnes represents a dominant point of view in certain philosophical circles. Those who subscribe to this perspective would tend to rule out the possibility that the innovation of philosophy and science owed a significant debt to the architects engaged in monumental temple building contemporaneous with the flourishing of Anaximander. Indeed the *supracelestial* view makes it impossible to take such an investigation seriously. The short article now being presented, part of this larger project of tracing out the origins of the earliest philosophers within their cultural context – and in this way re-appraising the very idea of *western rationality* – offers a challenge to Barnes's point of view.

This essay is divided into several sections. In (II) I offer a note concerning the cultural context in which trees, later replaced by stone columns in temple building, were regarded as sacred. In (III) I set out the problem of making a diagram or model of Anaximander's cosmos. In (IV) I outline the relevant fragments and testimony – for Anaximander's picture of the cosmos and its formation – from which a diagram or model might be constructed. Next, in (V) I invite the reader to reflect on the differences between plan and elevation perspectives: two ways of imagining. In (V.1), evidence for plan and elevation perspectives in ancient Egyptian architecture is examined, and the contributing influence is considered; in (V.2) evidence for plan and elevation perspectives in archaic Greek architecture, focusing on the Ionian evidence, is then taken up. In (VI), possible plan and elevation renderings of Anaximander's cosmos are displayed. And then, in (VII), some reflections on the consequences of this project are briefly considered.

II. SACRED TREES: A NOTE ON RELIGION AND SYMBOLISM IN ARCHAIC CULTURE: WHY MIGHT ANAXIMANDER HAVE BEEN INSPIRED TO ENVISION THE COSMOS IN THE IMAGE OF THE COLUMN-DRUM?

The argument that tries to make plausible the influence of the community of architects and their productions on the philosophical mentality of Anaximander must make plausible *why* Anaximander would have found the column-drum model appropriate for an understanding of the structure of the cosmos. In this short piece, I can only hope to sketch the outlines of the cultural context in which the symbolic significance of the column finds a special place. The stone columns were replacements for trees, and thus, it is to the symbolic significance of trees that we must turn our reflective attention. But, first of all, it must be said that ancient Greek religion was deeply symbolic, that sacrifice and prayer were central to its divergent practices, and that as Burkert put it, *sacrifice*, in particular, was the act of making something *sacred*.[14]

In the wide context of ancient Greek religion, trees were sacred; it might be argued that, within some cults, they were even more sacred than the altars that were constructed by their sides. Cult worship extends back into earlier times, but the general picture of the practices is reasonably clear. The tree, springing from the fertile earth, is an emanation of the divine. Altars were set up next to the sacred tree or grove – in Ionia, next to the Lygos tree for Samian Hera, and to the grove of Laurels sacred to Apollo Didymaios – and sacrifices were made to them. In Samos and Ephesus, the sacred precincts were swamps; all of nature comes alive in the presence of the watery and fertile land: frogs, small animals, birds come to feed, all kinds of vegetation flourish wildly. The presence of such living diversity is a sign of the presence of the god. Hera, not surprisingly, was presented in Samos as a plank of wood, no doubt, from the sacred Lygos tree.

In Ionia, the 6th century temple builders sought to enlarge the temple, whose roof was held up by tree-trunks. When this decision was made, the trees were no longer substantial enough to support the extended structure. The replacing stone architecture offered columns out of stone, thereby creating the experience of entering a petrified forest. The sacred trees were transformed but their symbolic sacred meaning could escape no one's notice. The column not only held up the roof but also symbolically traversed the space between man-made heaven and earth. The temple, symbolically imitated in microcosm the macrocosmic universe. This sentiment is echoed by Homer, Hesiod, Pindar, and others. Homer, in the *Odyssey*, reminds us of the story of Atlas, guardian protector of the great columns that keep heaven and earth apart.[15] The same story is alluded to by Hesiod in the *Theogony*.[16] And Aristotle mentions the story of Atlas in both the *Metaphysics* and in the *De caelo*.[17]

Pindar refers to the column that soars upward to heaven, connecting earth with the heavens rather than keeping us apart from it.[18] In a recent attempt at reconstructing Anaximenes' cosmos, McKirahan invites us to see an earth floating on a *column* of air.[19] And in the Myth of Er at the close of the *Republic*, Plato tells us of an adamantine column of light that unyieldingly holds the extremes of the heavens together, shining pure and bright, and around which the heavenly bodies orbit in concentric circles.[20]

All this testimony suggests a cultural context in which the column played a significant symbolic role, linking heaven and earth, and a significant sacred role in its unique capacity to touch both worlds, to hold them together. The column has a sacred symbolic and religious importance, connected first to the tree that embodied the divinity, and subsequently to its function in monumental temple building while never losing its original meaning. It is in this cultural context that Anaximander flourished and in which it is likely that he found both the column suitable for his cosmic imagination, and a community that could be expected to grasp immediately its cosmic appropriateness.

III. THE PROBLEM OF *IMAGINING* ANAXIMANDER'S COSMOS

More than thirty years ago, in important work by both Kahn and Sambursky, the opinion was expressed that Anaximander's cosmology permitted representation in a diagram. Kahn made the point that while it would be hopeless to draw a diagram of the poetic descriptions of the cosmos by Homer and Hesiod, "The characteristic view of [Anaximander's] earth [is] that it lends itself directly to geometric representation." And Kahn continues, "We can scarcely doubt that the Milesians were in fact accustomed to discuss such matters with the aid of diagrams or simple models."[21] Sambursky put forth the same sentiment when he declared that, "In the cosmology of Anaximander use was made for the first time of the scientific model as a means of description or as a method of explaining phenomena."[22] Although both scholars identify Anaximander as the first in a line of cosmologists whose geometrically-conceived model of the cosmos can be represented in a picture, neither offered us a possible rendition.

The problem of drawing a picture or making a model creates difficulties. When Anaximander *imagined* the cosmos, from what perspective or perspectives did he do so? Is it likely – and on what grounds – that some other part of his 6th century community inspired his imagination? Who else was engaged contemporaneously in drawing diagrams or model-making? Is there any light to be thrown on possible variations in perspective-representations with which Anaximander may have been familiar?

Presocratic studies, by the very fragmentary nature of the evidence, must be speculative.[23] In attempting to offer visual models, I am painfully aware of the degree to which guesswork enters into the formulation. Nevertheless, what I shall set out to do is to try to sketch a plausible case that Anaximander may likely have *imagined* the cosmos from more than one perspective or model, and that the community of architects/engineers[24] working contemporaneously on monumental stone temples to Hera in Samos, to Artemis in Ephesus, and to Apollo in Didyma, directly or indirectly stimulated his cosmic imagination.[25]

IV. PROSE FRAGMENTS AND TESTIMONY ON ANAXIMANDER'S COSMIC STRUCTURE AND ITS FORMATION

In the surviving prose fragments and testimonia, what claims can we reasonably accept about the image and formation of Anaximander's cosmos?

a) The shape of the earth is curved,[26] round, like the drum of a column (*kionos lithō paraplesion*);[27] the earth is cylindrical in shape, its depth is one-third of its width (= 3 x 1).[28]

b) From a conflict of opposites – of hot and cold – a sphere of flame was formed round the air surrounding the earth, like bark (*phloios*) around a tree (*dendron*), and when this was broken off and shut off in circles, the sun, moon, and stars were formed.[29] Thus, the sun, moon, and stars are circles of fire, encased in air, like certain pipe-like passages; what we identify as the sun, moon, and stars is the fire showing itself through holes in these fiery pipes,[30] as through the nozzle of a bellows (*prēstēros aulos*).[31] Each of these fiery circles is like the wheels of a chariot,[32] with its felloe hollow.[33]

c) The circle of the sun is 27 times that of the earth; the circle of the moon is 18 times the size, and presumably the circle of the stars is 9 times that of the earth.[34]

Sambursky emphasized that the model of revolving wheels and the fire appearing at the mouth of a forge are "perfect examples of technical analogy." The use of technical analogy indicates the "tremendous revolution in thought which took place in sixth-century Miletus."[35] And this assessment seems just right. However, the striking feature of this description of the cosmos that has been neglected in scholarly discussions is the architectural structure of the column drum, and the particular technical analogy on which Anaximander may have been drawing. If a plausible case can be framed that Anaximander thought through the cosmic structure by reflecting upon features peculiar to innovative techniques in column drum preparation, originating in the 6th-century monumental temples in Samos, Ephesus, and Didyma, we may have indication of unsuspected contributions from the community of architects/engineers to Anaximander's philosophical conceptions.

Hippolytus' testimony that Anaximander identified the earth as a *kioni litho* has not been an issue of contention, and the testimony is in accord with that offered by Pseudo-Plutarch. Further, column drum construction was a technical innovation in Ionia in the first half of the 6th century, the proportions of 3 x 1 would be broadly appropriate for the constructions themselves, and technical analogy, as Sambursky rightly observed, was characteristic of Anaximander's thought. The striking feature, however, of the picture projected by Anaximander, on the authority of the doxographical tradition, is that the earth is shaped like a column drum, three times as wide as it is deep, a "ratio which is analogous to the distances of the heavenly bodies."[36] The cosmos displays a structure analogous to the colum-drum earth; that is, the structure of the cosmos is expressed in terms of the structure of the column drum earth.

The case I am trying to sketch rests on two central points: (i) Anaximander *imagined* the cosmos in terms of a column drum earth; this image invites us to investigate a possible connection with the technology and design of archaic temple architecture underway in his own backyard; and (ii) Anaximander may likely have *imagined* the comsos from more than one point of view, and a possible source for inspiring these different imaginative perspectives might plausibly have been the architects.

V. TWO WAYS OF IMAGINING: PLAN VS. ELEVATION VIEWS

In the attempt to make plausible a connection between the efforts and productions of the community that undertook the task of planning and executing the monumental temple constructions – the architects/engineers – and Anaximander, the argument must secure some more specific claims. It seems quite possible that Anaximander imagined the cosmos from more than one perspective. And it might be that he did without ever having been impressed to do so by any acquaintance with the planning and construction of the architects. In the next section, I shall investigate the imaginative differences that become apparent when his cosmos is visualized in *plan* or *elevation*. And this argument can, I believe, stand separately in the absence of conclusive evidence. However, the case I want to sketch is the one that invites us to see a possible, and deeply interwoven, interaction within a social community that brought together *phusiologos*, like Anaximander, and *architekton*, like Rhoikos and Theodorus, Chersiphron and Metagenes.[37] The possible influence of the architects on the origins of early Greek philosophy has not been fully appreciated.

If Anaximander came to think of the cosmos from more than one perspective, he might have been inspired by the architects working in the second quarter of the sixth century. Is there any good reason to suppose that the archaic architects planned or executed their constructions in terms of plan and elevation perspectives? I believe there is reason to suppose that they distinguished between these two points of view; however, to what degree these perspectives were regular features of their work is difficult to establish. To investigate this case, I first turn to reflect on what we relevantly know about architectural drawings from Egypt, an important source of influence. I then turn to focus on the more conjectural case for sixth century Ionia.

V.1. Plan and Elevation Views in Ancient Egyptian Architecture

Why investigate the Egypt connection? From the mid-seventh century and following, the Ionians would surely have seen the monumental works of pharaonic architecture in dressed masonry, and more importantly could have learned from the busy architectural endeavors of Psamtik I, and his successors, how such buildings were erected. Around 660 B.C. the Egyptian pharaoh known to the Greeks as Psammetichos gained control over his country from the Assyrians with the help of mercenaries from Ionia and Caria. From that point on, close contact between Ionia and Egypt is evidenced in many forms not least of which was the establishment of the Greek trading colony at Naucratis in the late seventh century.[38]

The importance of Egyptian influence is part of the familiar discussion among historians of architecture.[39] Egyptian and archaic Greek temple architecture both depend on accurate megalithic masonry, and in the absence of monumental buildings in Ionia, Egypt would have offered ready examples of techniques for quarrying, transporting, and dressing huge monoliths. None of the emphasis on "influence," however, should undermine the equally strong case that Greek temple architecture developed in very different ways from that of the Egyptians. The relevant case here is that Ionian Greeks would have had the opportunity to see, first-hand, monumental temples, like the multi-columned temples at Karnak and Thebes, and Abydos and elsewhere. They would have been in a position to observe and marvel at the techniques of construction displayed in the ongoing building programs of Psamtik and his successors. And those who would finally have been entrusted to plan and supervise the archaic Heraion, Artemision, and Didymaion, would have had a chance to reflect on how the Egyptian architects/engineers *imagined* and produced their buildings.

How did the ancient Egyptian architects plan and execute their buildings? It is not easy to say with great confidence. There is no doubt that tremendous planning would have been required, but just how that planning was carried out is far more open to doubt. What we do know about building plans is detailed in the classic work on ancient Egyptian architecture by Clarke and Engelbach, a work that is still generally regarded as the standard. According to their work, the following preparations seem to have been undertaken prior to building:[40]

a) "Plans – perhaps models – of the proposed building had to be submitted to the king."

b) "Actual plans and models have been preserved."

c) "There were palace archives where plans of temples were preserved, since in one of the crypts at Dendera an inscription states that the plan of the temple was found, written in ancient characters, in the palace of King Pepi."

d) "Another passage relates that a restoration had been made by King Tuthmosis III after a plan had been found dating to the time of King Khufu."

e) "The Egyptians were able to draw an object from different aspects, showing side- and end-elevations, for example, but only one drawing has been preserved as definite proof."

f) "A truly sectional representation of a house, showing the contents of each storey, is known in the New Kingdom."

The front and side elevations of a shrine on papyrus, dating to the 18th dynasty – the New Kingdom (ca. 1580 - 1304 B.C.) – is pictured below:[41]

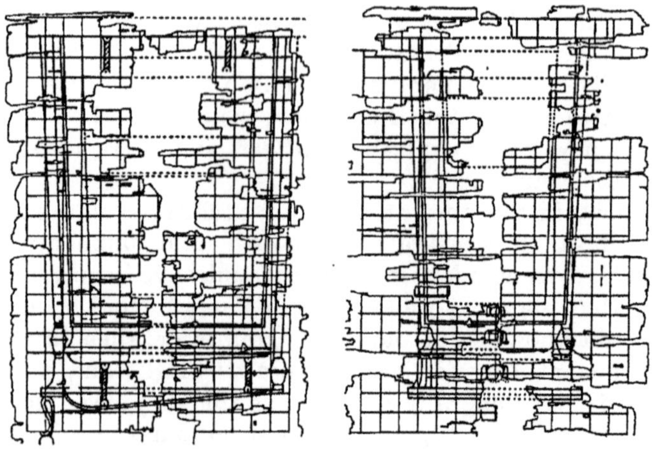

Next, we can reflect on an ancient plan, on papyrus, from the tomb of Rameses IV (ca. 1166 - 1160 B.C.).[42]

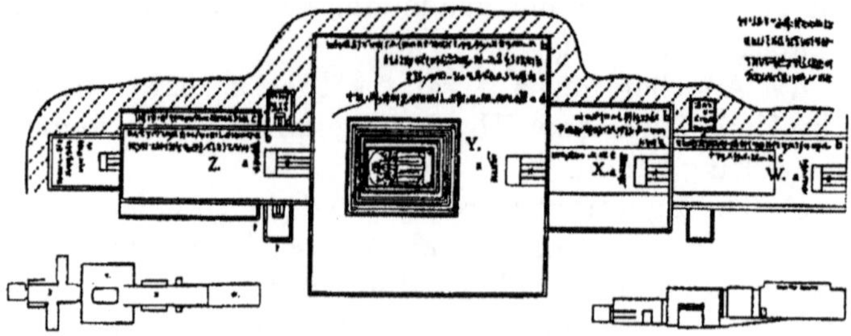

The next plan, on limestone, in what is probably the tomb of Rameses IX (ca. 1140-1123; according to Clarke and Engelbach "This plan should not be looked upon as the architect's original plan of the tomb, but rather a sketch-plan for the guidance of the workmen."[43]

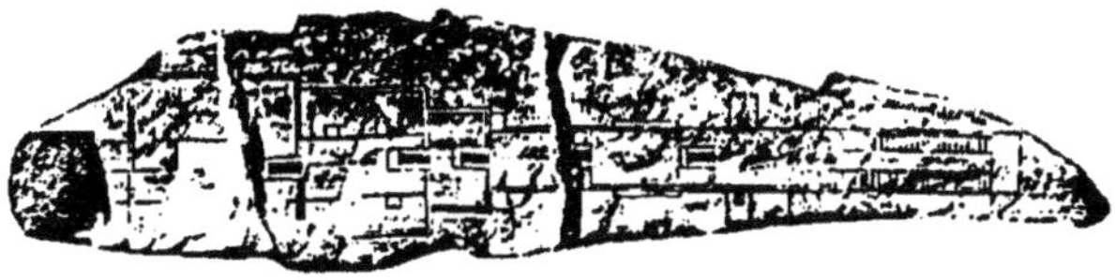

The diagrams just considered belong to the New Kingdom, but evidence can be produced pre-dating these by more than a millenium. Below, an architect's diagram defining a curve, by coordinates, probably dating to the third dynasty (ca. 2686 - 2613 B.C.) from Sakkara.[44] "At regular intervals (of 1 cubit each, though this is not stated explicitly) one should draw a perpendicular line of a stated length. The lengths are given in the cubit notation. . . . When the points at the ends of the lines are joined a curve is produced."[45]

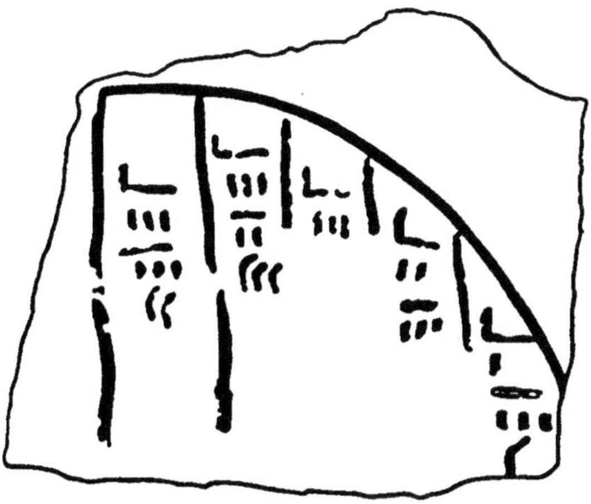

There are other important pieces of evidence, but for my purposes in this limited space these will suffice. From this evidence not much can be concluded with certainty about how the Egyptian architects worked. For there is nothing in this evidence or in the surviving models to show that the planning might not have been done directly on site at full scale, and that the sketches that survive served more as an informal aid to reflect upon than as a working plan at small scale.[46] The precise procedure by which the architects built must remain open to doubt. But the argument that I am advancing does not require a definitive statement on the relation of plans and models to building practice. What the argument must show, for my case, is that plan and elevation perspectives were commonly *imagined* by the ancient Egyptian architects, and that there is clear evidence that more than one perspective was regularly present in the minds of these architects when imagining, discussing, or erecting their buildings.

V.2. Plan and Elevation Views in Archaic Greek Architecture

A recent discovery by Haselberger at the temple of Apollo at Didyma revealed an entire archive of construction plans still in place on the temple walls themselves. Haselberger discovered, in finely etched lines, full-scale drawings for columns on the podium walls of the adytum. He discovered what he termed "geometrically pure paradigms" of the torus profile of a column base that proved the architect began with such a model and then refined and re-worked the material to achieve the desired effect. He found floor plans on the floor, and wall plans on the wall, full-size. And he concluded that, in the floor plans for example, if the tentative plans were accepted, they were copied in turn from layer to layer, while earlier steps were erased – polished over – as they went. The only reason, he supposed, why the plans remained at all was that the building never even neared completion.[47]

And not only in Didyma do we have such evidence of plan and elevation drawings. In the temple of Athena at Priene, Koenigs discovered a scaled-down sketch of its pediment, incised in a block that was later fitted into the building itself. And Hoepfner uncovered evidence of plans for a burial chamber that were drawn in red chalk on a segment of the temple of Artemis in Sardis.[48] These kinds of evidence indicate clearly that the architects/engineers relied on drawings, sometimes in plan and sometimes in elevation, in the process of their construction. Although the Ionian evidence by

Haselberger, Koenigs, and Hoepfner belongs to the late classical and early Hellenistic periods, it has laid to rest the question of whether or not the architects made plans.[49] But, the problem that still remains is what to make of the earlier constructions whose technological display would certainly have first amazed the Ionian populations.[50]

The evidence for architectural planning in the archaic period has become much clearer since Kienast's work on the so-called Temple D at the Samian Heraion dating to the late sixth century. In earlier discussions, the evidence for architectural construction was less clear. Evidence for early clay models of temples and houses was known, but the architectural use of these models was and is still in doubt.[51] The likeliness that models, in clay or other materials, exhibiting an elevation view of the proposed temple were presented as part of a strategy for securing patrons seems great. After all, it is difficult to accept that patrons would agree to fund a project at an exorbitant cost for so many years without a model exhibiting the finished temple. But, in the absence of evidence to the contrary, the supposition that "plans" preceded the constructions themselves was dismissed as fanciful.[52]

But the understanding of how the Greek architects worked has become clearer in the last decade, although far from clear. Haselberger's discovery, of course, startled those who doubted extensive planning in the form of scale drawings. And then Kienast published his piece on the so-called Temple D. Kienast knew that many of the buildings indicated, by scratching and other markings especially at corners and shafts, that planning had been undertaken at the site prior to construction. But, at the so-called Temple "D," he discovered evidence of a complete groundplan (*eine vollständige Grundrißzeichnung*), marked out in red, transcribed directly to the construction foundation, 1:1.[53] This proves, in his estimation, that the archaic architects in Ionia imagined their constructions in plan view, transcribed the plan to the construction site itself, and erected the building on just those lines in which the plan consisted. In the diagram below, after Kienast, the heavy black lines indicate the red lines found in the sixth century construction:

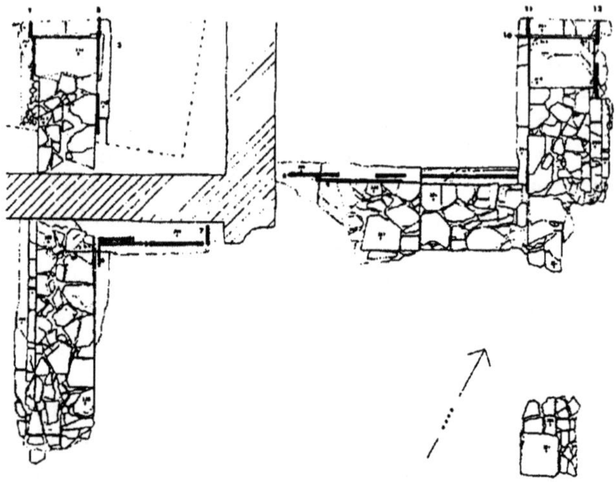

In another project, Kienast published a definitive work on the planning and execution of the tunnel of Eupalinos on Samos.[54] Although the tunnel construction belongs to the period just after Anaximander's "publication" – roughly between 540 and 522 B.C. – Kienast's case is that it displays just the kind of techniques in planning that were available. To put the matter simply, the construction could not have been effected without a plan model. The tunnel is more or less 1040 meters long, driven separately from two sides; it runs some 400 meters in the south end and just over 600 meters in the north end. Kienast contends that the hill was staked out in order to determine the length of, and straight line for, the proposed tunnel. And he discovered at least five different marking systems in the tunnel – ancient survey markings – painted in red, one of which led him to speculate that Eupalinos invented his own tunnel-measure.[55] But the great and unexpected difficulties for the architect, he discovered, arose in the north end when Eupalinos and coworkers discovered loose rocks and the occurence of a great amount of natural ground water. Eupalinos decided to leave the straight line in the north end and chose to turn northeast, that is, into rather than away from the hill towards the sea. When he abandoned the proposed straight line, the technique of staking out the hill lost its effectiveness; then, Eupalinos would have had to rely on a variety of plans to insure that the tunnel halves would meet as initially intended. Kienast's reconstruction of the architect's technique claims that while the survey of the original plan centered on the straight line that was staked out across the mountain crest, a re-adjustment of this plan was necessary to handle unexpected deviations and yet still arrive at the anticipated meeting point. Without recourse to such plans, the architect would not be able to determine where he was in the hill and so not be able to control the project.

Although the Eupalinion post-dates Anaximander's prose writing, it does not do so by a margin of time that undermines our understanding of the architectural techniques that could have been known to Anaximander. Eupalinos' construction has no comparison in the archaic Greek world, let alone Ionia. The applied geometry illustrated in the construction techniques would have already been vindicated in other enterprises in order to encourage the Samians to undertake an unparalleled project lasting more than a decade. Eupalinos comes from Megara on the mainland; the architects – often comprising an itinerant community – provided one means for the collection and dissemination of building technologies throughout Greece. Kienast's work on the Eupalinion, and the so-called Temple D, helps us to understand more clearly how the architects faced up to serious difficulties in their constructions. His work offers us the archaic evidence that the architects working in Ionia imagined, and set out, their constructions in plan prior to the constructions themselves.

Before turning to try to sketch Anaximander's cosmos, one more set of illustrations is in order. Just in case the idea of *plan* vs. *elevation* view is still not clear, these differing views are presented pictorially for the so-called Rhoikos/Theodorus temple to Hera in Samos (begun around 575 B.C.),[56] and the so-called Chersiphron/Metagenes temple to Artemis in Ephesus (begun around 560 B.C.). The overall structure of archaic Ionic temples is the same; they are roughly 1 unit in height, by 2 units in width, by 3 units in length. In the reconstructions for the Samian Heraion proposed by the excavators, the plan is roughly 172.2 feet in width and 344.4 feet in length. The reconstructed elevation view presents a building whose columns are more than 50 feet in height, and with the entablature, probably extended an additional 35 to 40 feet:[57]

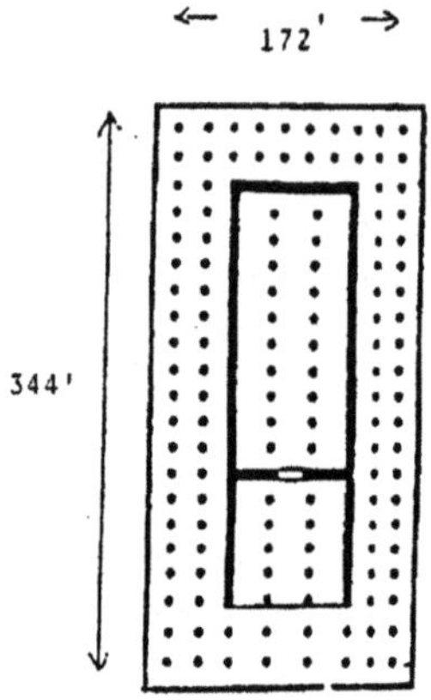

PLAN
(after Kyrieleis)

ELEVATION
POSSIBLE RECONSTRUCTION
OF THE RHOIKOS/THEODORUS
TEMPLE OF HERA c. 575 b.c.
(after Tomlinson)

Along the same lines, according to the excavator, the archaic Artemision identified with the architects Chersiphron/Metagenes, was roughly 180 feet in width and 377 feet in length. The reconstructed elevation displays a building with 60 foot columns whose entablature extended an additional 30 to 40 feet:[58]

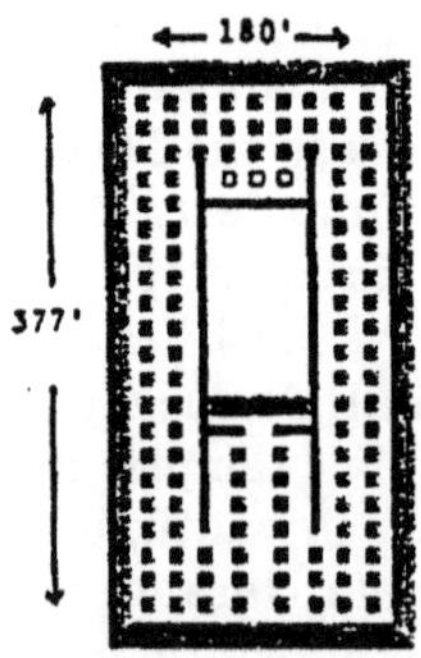

PLAN
(after Tomlinson)

ELEVATION
POSSIBLE RECONSTRUCTION OF
THE CHERSIPHRON/METAGENES
TEMPLE OF ARTEMIS IN EPHESUS
c. 560 b.c.
(after Grant)

VI. PLAN AND ELEVATION VIEWS OF ANAXIMANDER'S COSMOS

Precisely how the archaic architects built must remain open to doubt. The tradition of ancient Egyptian architecture offers evidence that both plan and elevation views were part of the consciousness of its builders. The archaic Greek architects, on the contrary, were not the product of a long and impressive tradition. In fact, the evidence suggests that there were no quarries in operation in Greece from the period of the fall of Mycenae until roughly 700 B.C.,[59] and hence no truly monumental architecture.[60] With dozens of generations engaged in no monumental construction, the idea and techniques for monumental temple building had to be imported, and inspiration from Egypt is persuasive.

The evidence from archaic Greece is less conclusive. But, it seems reasonable to suppose that both plan and elevation views were also part of the consciousness of the archaic architects.[61] The elevation view or model would have been particularly important in securing patrons; otherwise, we must believe that a project requiring hundreds of men for thousands of days would have been funded blindly. The plan view, on the other hand, is always the view of the builder at the earliest stages of construction. The higher levels, of course, cannot be constructed without the lower levels in place, and once in place the lower levels cannot be modified at all in light of what follows. It is worth emphasizing that mistakes made from the start cannot be corrected and will be ruinously expensive. It is for these reasons that the architect building on monumental scale must have a technique of design that will allow him to visualize the completed building with sufficient accuracy so that the lower parts will be in accord with the upper parts and the finished building will achieve the desired appearance without collapsing under its own weight.[62] To achieve this aim, the plan and elevation views must surely have been part of the consciousness of the archaic Greek architects/engineers.

We must also keep in mind the changing environment of sixth century Ionia. There were many ingredients that, no doubt, contributed to Anaximander's mentality but the drastic change in the landscape, announced by monumental temple building, should not be underestimated. A key architectural feature in the archaic temples to Hera in Samos, Artemis in Ephesus, and Apollo in Didyma, all underway prior to the "publication" of Anaximander's book,[63] was the column construction.[64] Unlike the earlier buildings which focused attention on a house in front of the altar, the profusion of columns consciously hid the inner cella. The Samians, Ephesians, or Milesians who approached these buildings found themselves overwhelmed by these *thaumata*,[65] these sources of awe and wonder. Approaching the great temples, the Ionians met a veritable forest of columns; the experience must surely have been one of gazing into a petrified forest. The Ionian structure, unlike the Doric, is much livelier and more delicate. The columns spring upwards from a platform barely above the ground and reach upwards to the sky as if some vegetation flourishing in the marshy fields sacred to Hera and Artemis. In the earlier temples, the columns were made of tree trunks and reached toward the sky quite naturally; in the monumental innovations, stone columns replaced the tree trunks that limited the size of the building. The point of emphasis is that the double peristyle was central to the stunning outward appearance, that these enormous temples astonished the Ionians, and I am supposing that Anaximander was among those deeply impressed.

Is there any good reason to suppose that Anaximander envisioned his cosmos from more than one perspective? The case that he did, of course, rests on conjecture because no diagrams or models attributed to Anaximander survive. Learned scholars like Kahn and Sambursky supposed he did make use of diagrams or models but *imagining* those pictures must always invite doubt.

According to a reliable tradition, traceable through Agathemerus[66] and Strabo,[67] Anaximander is credited with drawing the first Greek map of the inhabited world on a tablet.[68] Such a map could have been attempted from the reports of seafaring people who passed through cosmopolitan Miletus. How much the map depended upon his reflections on the heavens cannot be determined. But, Anaximander is also credited with setting up a seasonal sundial in Sparta.[69] If true, he would have focused on the rising and setting of the sun on the summer solstice, the winter solstice, and the equinocts; these cosmical events, as Heidel pointed out,[70] framed the three-point coordinate system of the Greek map. The seasonal sundial and the map of the inhabited earth would seem to have been connected. In any case, a map of the earth, given the fact that Anaximander believed the earth to be a *flat* cylinder, would have had to be a *plan* view. Below is a possible rendition of Anaximander's map, by Robinson.[71] I have inscribed the map on a column-drum in accordance with the testimony.

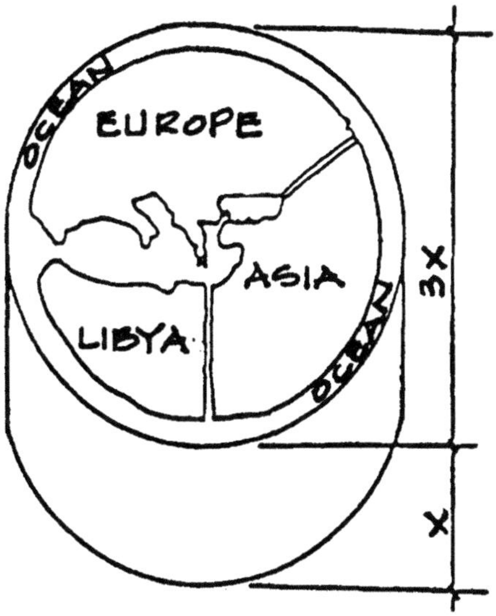

For my purposes here, I am not interested in entering into a debate about the details of the map. My only point is that such a conception lends itself readily to a *plan* view rather than an elevation.

Anaximander's cosmos is geometrical; this feature, as Kahn and Sambursky pointed out, makes it amenable to graphic representation. According to the cosmology, from a conflict of opposites – of hot and cold – a sphere of flame was formed round the air surrounding the earth, like bark (*phloios*) around a tree (*dendron*); and when this was broken off and shut off in circles, the sun, moon, and stars were formed. Below, on the left is an attempt to render a picture of the flame of fire and the inner rings into which it is somehow broken off. On the right is a simple rendition of a cross-section of a tree that explicitly serves as the metaphor.

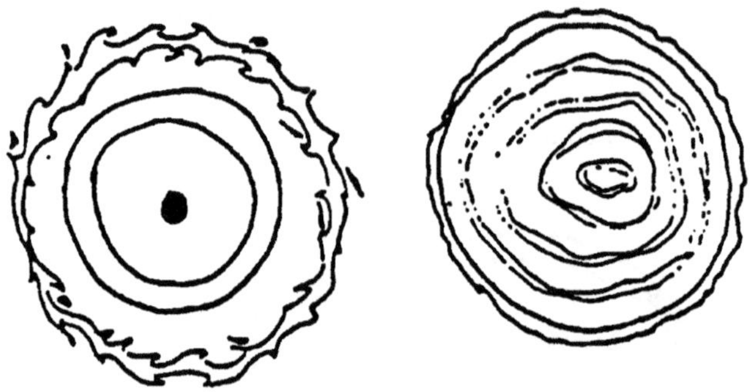

Now, according to the cosmology, these inner rings are made of fire, encased in air, and what we identify as the sun, moon, and stars is the fire showing itself through holes in these fiery pipes as through the nozzle of the blacksmith's bellows. Below, then, is an attempt to render Anaximander's geometrical cosmos in a *plan* view:[72]

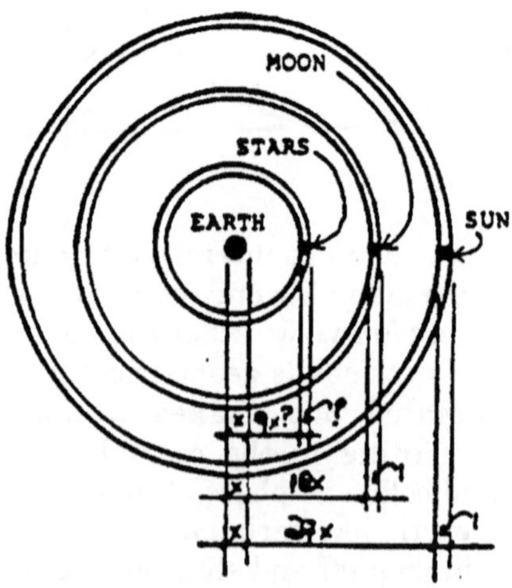

With these renditions in mind, it is now time to return to the architectural discussion. Anaximander, on the authority of Hippolytus, identified the shape of the earth with a column drum. So, it is appropriate to take a closer look at some drums that he might have seen in order to determine if there was anything else about the column drum that seemed to suggest itself as illustrative of the cosmic model. Column drum construction was new to Ionia in the sixth century B.C. and with it came a new architectural technique for preparing the drums that would constitute columns fifty feet, or more, in height. That technique is displayed on archaic drums from the Ionian temples; the technique is called *anathyrōsis*.

The term *anathyrōsis* is identified with a labor-saving device by which contact between two blocks was obtained by dressing only the edge around the tops and edges.[73] The procedure was usually employed on the vertical faces between two blocks.[74] In the usual masonry technique, the horizontal faces of the stone blocks were completely dressed to a plane, but the vertical faces could be fit well without having to dress the entire surface.[75] In the development of monumental building, the fit gained from *edge anathyrōsis* proved not sufficiently precise as the blocks became increasingly massive. The next step was to dress the vertical sides with a band around all the edges, not just the top and side, and this technique is sometimes referred to as *band anathyrōsis*. The *anathyrōsis* technique – which derives its name because the effect is something like the frame of a door (*thyra*) – in the case of edge *anathyrosis*, was a solution to the problem of precisely fitting one block to the next without mortar.[76]

In column construction, band *anathyrōsis* is already in evidence from the mid-seventh century;[77] column bases were prepared with a smooth band running around the circumference of the horizontal joint face. The inner part of the horizontal surface was left rough but slightly sunken creating a concave surface. In the sixth century, in addition to the band *anathyrōsis* preparation of the column drums, another technique was employed for lowering the drum into place without chipping the sides. This device became known as the *empolion*;[78] it consisted of a square hole in the center of the drum through which a wooden pivot would be fit. The drum could be lowered into place directly centered on the lower drum. In order to determine precisely the center of the drum, additional concentric circles were constructed on the drum face. If the center was not placed exactly, the effort and cost of re-shaping the drum at the extremities would have been prohibitive. Below is a diagram of drum *anathyrōsis*; in this case, the drum has been fluted.

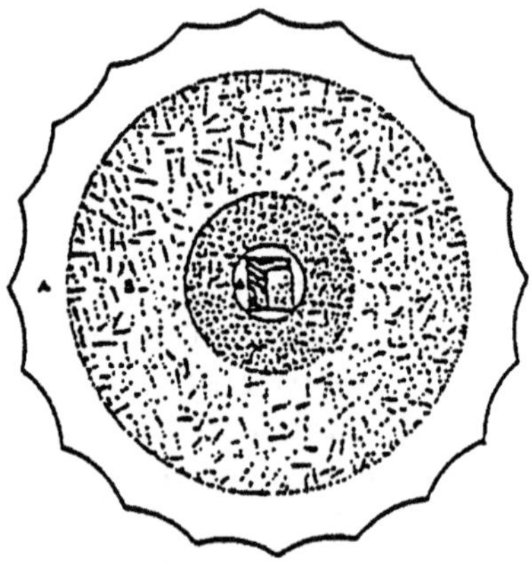

It is apparent that the horizontal face of the column drum exhibiting *anathyrosis* bears a striking resemblance to a *plan* model rendition of Anaximander's cosmos. By "plan" I mean the kind of perspective that one would assume if one envisioned a horizontal cross-section of the cosmos. The argument here is not that the drum face and Anaximander's cosmos display a one-to-one correspondence with his cosmos. Rather, Anaximander's identification of the earth with a column drum, whose 3 to 1 ratio is analogous to the distances of the heavenly bodies, seems more than fortuitous. In seeing a column drum perhaps he was inspired to *imagine* the cosmos, from one point of view. In the plan view, the fluting might be construed as a visual presentation of the ring of fire; the concentric bands effected by the *anathyrosis* technique suggest the wheels of the heavenly bodies.

Next, we turn to consider Anaximander's cosmos in an *elevation* view. Is there any reason to suppose that he would have thought it through from this perspective? Of course, we can ask how such a model would appear even if he had not done so. But, it is hard to make sense of even the outlines of Anaximander's picture without imagining it in elevation. This case is all the more compelling if we accept the attribution to him of inventing or setting up a seasonal sundial. No astronomical expertise is required to notice that, in Miletus, the sun is higher in the sky during the spring and summer months and lower in the sky during the late fall and winter months. Even the simplest picture of stellar regularities must account for the changing *elevation* of the sun in the course of a year.

Anaximander's account of the sun, moon, and stars requires that we imagine a series of revolving wheels, and consequently the mechanism that accounts for the change in their altitude.[79] Heath attempted to grasp Anaximander's picture, and his illustrative drawing is a good place to start. In his picture, the wheels of the moon and stars are omitted.[80]

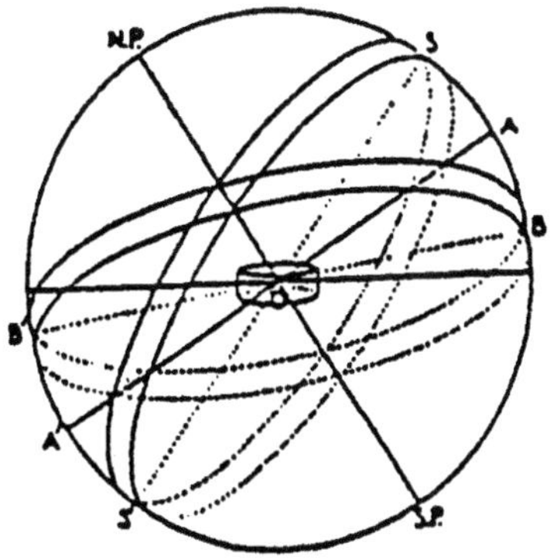

A more promising rendition has been proposed by Couprie.[81] His suggestion is as ingenious as it is conjectural. He invites us to imagine three concentric and telescoping *cylinders*. The holes out of which their fire shines turn around with their respective cylinders. The rings or wheels slide up and down on these invisible cylinders. The sun's wheel is a height of 47°, that is, two times the inclination of the ecliptic; this distance will suffice to account for the winter and summer solstices. In order to account for the monthly path of the moon through the zodiac, Couprie assigns a height to the moon's wheel of 57°. Inside these two cylinders is another cylinder of infinite length that contains the stars. And finally, inside the star cylinder is the flat cylindrical earth.

How shall we account for the mechanism that regulates the changing altitude of these wheels? No clear explanation is offered; Anaximander's picture describes rather than explains the phenomena.[82] But Couprie's ingenuity deserves our reflection. The picture he imagines, on Anaximander's behalf, follows through on the idea that the big cosmic structure is an expression of the small earthly structure. The ratio of the distances to the heavenly wheels is analogous to the ratio of the width and depth of the earthly cylinder. The cosmos is envisioned in terms of the earthly *cylinder*; the heavenly wheels are analogously interpreted as parts of cylinders.

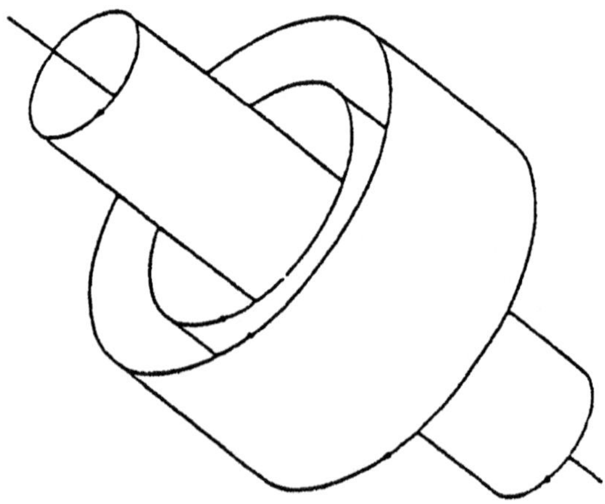

VII. EPILOGUE: TWO CONSEQUENCES

VII.1

An important problem in understanding Anaximander's cosmos was recently resuscitated by Furley.[83] The problem arises when we try to make sense of Aristotle's testimony in the *De caelo*. In one of the very few passages in which Anaximander is identified by name, Aristotle singles him out as being among the ancients who held that the earth remains at rest because it is in equilibrium (*homoiotēta*).[84] The earth, says Aristotle about Anaximander, is at rest in the center and does not move up or down, or to the sides, because it is equally related to the extremes (*homoios pros ta eschata exon*), and thereby has no reason to move one way or the other. Furley follows the problem as raised by Heidel,[85] and then explored in greater depth by

Robinson.[86] Only a spherical earth – not a flat and cylindrical earth – is equally related to the extremes. So, Aristotle, according to Furley, has somehow got it wrong. Instead, Furley defends the reasoning offered by Simplicius that Anaximander's earth remains at rest in the center because it floats on air.[87] In keeping with the Milesian tradition of a Thales who seems to have believed that the earth floats on water, and an Anaximenes who seems to have believed that the earth floats on a column of air, so Anaximander, like Anaximenes, held that the earth floats on air. The reason Anaximander held the earth to be a flat disc, according to Furley, is so that it could remain aloft.

Aristotle, no doubt, may have gotten it wrong, as he has in other cases involving the Presocratics. But suppose he didn't get it wrong, that he had in front of him a copy of Anaximander's book, or a summary from one of his students, when he wrote that passage in the *De caelo*. How could we reconcile Aristotle's testimony against the charge that only a spherical earth could be equidistant from all extremes? The approach I have proposed offers a resolution without having to suppose still another possibility, namely, that Aristotle is reporting accurately and it is Anaximander's image itself that was ill-conceived.

If Anaximander had imagined the cosmos from plan and elevation perspectives, the way in which the earth would be situated would not be the same. In each view, like that of the temples, the harmony and order would be perceived differently.[88] If one takes for granted that Anaximander's picture is strictly an elevation view, Furley's objection is hard to discount. But what requires us to suppose that Anaximander's model was exclusively an elevation? Had he envisioned the cosmos, as he might likely have drawn the map of the inhabited earth on a tablet, in a *plan* view, Aristotle's testimony could be preserved. For then, in plan view (that is, a horizontal cross-section), the round earth *is* equally related to the extremes. In the plan view, the earth is in equilibrium in the cosmos; it is equidistant from the heavenly wheels that stand in geometric proportions to the column-drum earth. Thus, an additional consequence of accepting this multi-planned interpretation of Anaximander's *imagination* is to preserve the testimony of Aristotle. And this is not unimportant, for it is difficult to accept that in one of the four times that Aristotle singles out Anaximander by name, he has simply got it wrong.

VII.2

Finally, the idea that the community of architects influenced the philosophical conceptions of Anaximander is surprising to the degree that we have embraced, perhaps unconsciously, Barnes's *supracelestial* perspective. To the degree that we have come to suppose the western tradition of rationality consists in the triumph of the mind over the body and senses, to that degree the thesis that the architects, directly or indirectly through their productions, inspired Anaximander's cosmical imagination, will be surprising. To take the thesis seriously, we must be prepared to rethink what is relevant to an understanding of philosophy, and to rethink the role that the imagination contributes to it. We must ask, anew: Are images essential to thought and rationality?

Traditional studies on rationality, like those of Descartes and Kant, routinely distinguished between concepts and images; whereas concepts were regarded as purely rational, images could claim a rational character only derivatively. This separation of images from rationality has been a consequence of traditional approaches to imagination that have proceeded either by regarding the imagination (i) as merely tied to the body, in a mechanistic way, generating images out of sense data, or (ii) as completely free, undisciplined and unfettered, and in this sense an expression of radical creativity. In the first case, the imagination is closely identified with the bodily aspect of experience, in the second case with mental activity that confounds rules; the familiar vision of rationality as the triumph of the rule-governed mind over the body and senses militates against treatments of the imagination as central to thought and rationality.

In recent studies, however, by those like Mark Johnson,[89] Mark Rollins,[90] and Roger Shepard,[91] the nature of rationality has undergone a reappraisal and the role that imagination contributes to it has been vastly transformed. According to these new studies, imagination is now being seen in the cognitive patterns of all of our concepts; indeed, patterns of understanding are now being studied as patterns of imagination. Consequently, the new view that is emerging is one that envisages imagination to be inextricably bound to an understanding of rationality rather than as a separate dimension of experience. Imagination, then, does not come into play only in moments of whimsy and radical creativity, but rather imagination becomes the locus for meaning, understanding and reasoning.

Imagination is now being discussed in terms of patterns that are shared by people; that is, rather than being an idiosyncratic and private expression, the imagination is now being investigated in terms of the commonalities that are displayed in the structures of understanding and metaphor. Thus, according to the lead of recent studies, to say that we are *rational animals* is to say that we are *imaginative animals*. Along with this new perspective has come a reappraisal of the traditional bifurcations between the rational and the bodily, between science and art. With the collapse of this strict dichotomy, the patterns of thought in science are being increasingly examined as expressions of the imaginative domain of lived experience. This new approach shows up clearly in studies in *Science, Technology, and Society* [STS] where science is investigated as a cultural practice.

Recent work in the history and philosophy of science and in STS have focused on investigations of scientific practices as embedded within a culture.[92] Rejecting the positivistic approaches that supposed an ultimately objective model of how the world is, and the sweeping generalizations towards which positivism strives, the new approaches have emphasized *case studies* and pursued objectivity within a local, rather than global, framework. These ground-breaking, historically-based, case studies have not only helped us to understand better precisely what the practitioners of science believed they were doing but also have assisted in the general project of reappraising the nature of rationality with which "science" has been familiarly aligned. One important consequence has been to open up the discussion of science and scientific practice to a consideration of its imaginative dimension.

The short study of Anaximander I have just presented tries to make sense of his cosmic imagination. But this study, as I have envisioned it, is not just about our ability to imagine ancient cosmological models, but rather to see that the origins of Greek philosophical rationality cannot be properly understood independent of this cultural embeddedness. My project on Anaximander, of which this is a part, seeks to show that an understanding of western rationality requires us to think through the material world rather than abandon it and any sense knowledge of it. The material world, in all its cultural breadth, is *indispensable* to an understanding of ancient Greek philosophical rationality and to the rationality that we philosophers are seeking to grasp. Thus, the project is not just to understand cosmological models but to show a conception of philosophy in the context of model-making and the imagination it presupposes. For in the absence of adequate astronomical instruments or theories, Anaximander *imagined* a geometrically-modeled cosmos. The "rationality" that his model exhibited was one whose warp and woof were the fabric of his Ionian techno-culture.

Southern Illinois University at Carbondale

NOTES

Earlier drafts of this paper were presented to the International meeting of the Society for Philosophy and Technology in Mayaguez, Puerto Rico in March 1991, and subsequently to the Society for Ancient Greek Philosophy at the annual meeting of the American Philological Association, December 1992.

[1] Aristotle, *Metaphysics*, 981b17ff; cf. also *Politics*, 1292b25ff, 1293a1ff, 1320a17ff.

[2] George Sarton, 1953, pp. 160-177. This thesis has also been defended by Charles Singer.

[3] Goody and Watt, 1968, pp. 27-68.

[4] Farrington, 1953, ch.2.

[5] Lloyd, 1979, pp. 234-264. I am following his exegesis of the competing hypotheses.

[6] Gernet, 1981, pp. 352-364.

[7] Vernant, 1982, pp. 285ff.

[8] Vidal-Naquet, 1967, pp. 51-58.

[9] Detienne, 1967, pp. 99ff.

[10] Lloyd, 1979, p. 248.

[11] Ibid. p. 248. That the spheres of law and justice provide important models for cosmic order and justice has been defended for a long time. Cf. Hirzel, 1907; H. Gomperz, 1943, pp. 161-176; L. Gernet, 1955; G. Vlastos, 1947, pp. 156-178, and again, 1975.

[12] It is important to note that Lloyd's position has been subjected to criticism by both Frischer, 1982, and Hurwit, 1985. The argument for the importance of participatory democracy to the development of Greek philosophy is partially undermined by the prevailing conditions in Ionia in

the early 6th century B.C. where tyranny, not democratic practices, predominated. Lloyd anticipates some of these objections, 1979, pp. 260-264. The argument I outline tries to take this criticism into consideration.

13 Barnes, 1979, vol. I, p. x. Cf. also his response to the criticism his work generated on just this point, 1982, p. xvi. Barnes modifies his rhetoric as a result of the criticism but his position is substantively unchanged: "Some critics, indeed, have accused me of being anti-historical, and their accusation has some point: I made one or two naughty remarks about history, and I occasionally flirted with anachronistic interpretations of Presocratic views. For all that, the book is a sort of history: it recounts past thoughts, and its heroes are long dead. In speaking slightingly of history I had two specific things in mind – studies of the 'background' (economic, social, political) against which the Presocratics wrote, and studies of the network of 'influences' within which they carried out their researches. For I doubt the pertinence of such background to our understanding of early Greek thought" (my italics).

14 Burkert, 1983, p.3; sacrifice is pre-eminently essential to the practice of ancient Greek religion, and Burkert details, on the authority of Homer and elsewhere, the ordinary ritual in which the sacrificial animal has its throat slashed, the blood splattered on the altar, the parts dismembered and cooked, and finally the bones set out and symbolically reconstructed.

15 Homer, *Odyssey*, I. 53ff.

16 Hesiod, *Theogony*, 517ff.

17 Aristotle, *Metaphysics*, 1023a19ff, and *De caelo* 284a20ff, where his concern is to reject the thesis that heavenly bodies are made of heavy material. It is also noteworthy that the tradition on Atlas has more than one rendition; the early story seems to identify him as the guardian of the columns while later he seems to take over their function himself.

18 Pindar, *Pythian Odes*, I. 39ff.

19 McKirahan, 1994, ch. 3. If one accepts Aristotle's testimony at *De caelo*, 294 b 13, that the flat, disc-like earth floats on air "covering it like a lid," then the image of a column of air readily presents itself. McKirahan offers a pictorial representation of this image.

20 Plato, *Republic*, X, 616 B ff. A Pythagorean vision.

21 Kahn, 1960, p. 82.

22 Sambursky, 1956, pp. 13-14.

23 There are those, like Dicks, 1970, p. 43-45, and 1966, pp. 26-40, for instance, who believe the tertiary evidence is so unreliable that nothing can be reasonably concluded. Cf. the rebuttal by Kahn, 1970, pp. 99-116.

24 There is no technical term in Greek for "engineer." Herodotus (3.60, cf. also 4. 87) uses the term architekton when referring to Eupalinos of Megara who supervised the construction of the tunnel/water-channel in

Samos, around 530 B.C. For a discussion of the architect, cf. Coulton, 1977, ch. 1. For the distinction between architect and engineer, cf. the thoughtful essay by Holloway, 1969, p. 286ff. Some person or persons were chiefly concerned with the overall design of the building and its detailed parts; this person(s) may be called the "architect": some person or persons were responsible for executing the construction – quarrying, transporting, placing, and dressing the stone; this person(s) may be called the "engineer." It may be that the architekton was responsible for all aspects of the construction and the various tasks were shared by teams of architects.

25 I want to be clear from the start that I am not arguing Farrington's thesis, 1949/1961, that technology proved to be a sufficient condition in accounting for the rise of Greek philosophy. Lloyd, 1979, p. 235, correctly criticized Farrington's thesis. If technology were sufficient then Egypt and Mesopotamia should have witnessed the birth of philosophy for they excelled over the Greeks in technological mastery. My thesis only entails the view that certain contributing aspects of technology have not been fully appreciated. However, if we make a list of the achievements, real or imagined, attributed to Thales, Anaximander, Rhoikos/Theodorus and Chersiphron/Metagenes, and place them side by side, the kinship is striking. Thales is credited with predicting some sort of stellar anomaly, generally spoken of as the prediction of a solar eclipse, the measurement of the height of a pyramid, the measurement of the distance of a ship at sea, diverting the river Halys for Croesus' army, among other things. Anaximander is credited with the first geometrical model of the cosmos, the first Greek map of the inhabited earth, and the first seasonal sundial. Theodorus is credited with inventing or introducing into Greece the set-square, the level, the rule, the key, the lathe, diverting the river Imbrasus in order to set the platform for the Heraion, and a new technique for casting life-size bronze statues. Chersiphron is credited with inventing a device for moving huge monoliths, and his son Metagenes gained esteem for developing that technique for the delivery of monolithic architraves. The broad family resemblance of these achievements is a kind of applied geometry with technological innovation; the kinship suggests a community of common interests. Cf. also Snodgrass, 1980, pp. 142ff.

26 I follow Kahn, 1960, pp. 55-56. If we emend *guros* for *hugron*, the word "curved" must be interpreted to mean "concave" rather than "convex." (Cf. also Burnet, 1945, p. 65, n.1.) The familiar Ionian doctrine is that men live in a hollow of the earth, that is, the Mediterranean basin. As Kahn noted, this is also the teaching of Anaxagoras, Archelaus, and Democritus.

27 Diels-Kranz [DK], 12B5. Hippolytus, Ref.1,6,3. On the reliability of Hippolytus on Anaximander, cf. Kahn, 1960, p. 15: "All of the information which this author [Hippolytus] gives us concerning Anaximander (with the exception of his date . . .) comes from Theophrastus and from no other

source. He is drawing on an epitome in which information spread throughout the sixteen or eighteen books of the *Phys. Opin.* had been grouped under the names of various thinkers. The account of doctrines has been abbreviated. . . . But no basic blunders mar these excerpts." Hippolytus is, along with Simplicius, our best surviving source for Anaximander.

28 DK 12A10. Ps. Plutarch, *Strom.* 2. Cf. Kirk-Raven, 1957, p. 134; Guthrie, 1962, I, p. 95. Cf. also the interesting article by O'Brien, 1967, esp. pp. 424-425; he points out the difficulty of the expression *exein de* (sc. *tēn gēn*) *tosouton bathos hoson an eie triton pros to platos* and suggests that its meaning might be that the height is three times the size of its diameter, not one-third. According to the more widely accepted interpretation, that the earth's diameter is three times its height, the earth would be more likely to float on air as a reasonably flat disk. However, if one accepts that the earth is held aloft *dia ten homoion panton apostasin* then it does not matter whether the earth is conceived as a longer cylinder rather than as a flatter disk.

29 DK 12A10. Ps. Plutarch, *Strom.* 2. The proposed order that the wheel of the stars is closer than the moon and sun is unusual. Kahn, 1960, p. 90, proposed a completely "rational" explanation: where there is more fire, and hence brighter, the wheel is more distant; thus if the stars were brighter, they would be further, but they are not brighter, therefore they are not further. Burkert, 1963, suggested Zoroastrian influence by pointing to passages in the Avesta that offered precisely the same cosmic arrangement where the stars were closer than the sun and moon. West, 1971, p. 109, agreed with Burkert and concluded that there were two main components of Anaximander's vision: "a native tradition of materialist meteorology and physics, and an oriental tradition of metaphysical speculation." West's general conclusion was that, p. 97: "Anaximander's conceptions cannot be derived from Greek antecedents, and to suppose that they chanced to burgeon his mind without antecedents, at the very moment when the Persians were knocking at Ionian doors, would be as preposterous as it was pointless."

30 DK 12A11. Hippolytus, Ref. 1,6, 4-5.

31 DK 12B4. Aetius II, 20, 1.

32 DK 12A22, Aetius, II, 25, 1, and DK 12A21, Aetius, II 20, 1: *hamarteiō troxō.*

33 Ibid., *tēn hapsida exonta koilēn.*

34 DK 12A11. Hippolytus, Ref. 1, 6, 4-5; and DK 12A21, Aetius, II, 21, 1. I follow the discussions in Kirk-Raven, 1957, pp. 134-135, and West, 1972, ch. 3. But, cf. also O'Brien, 1967, pp. 423-432, who calculates differently and cannot be easily dismissed. He wonders about the diameters of the fiery wheels (not simply to be confused with the apertures that appear on

the inside face of them) and whether the distances to the stars, moon, and sun are measured in terms of 9, 18, and 27 earth diameters or radii. The difference changes the numbers but not the proportions. The importance of O'Brien's article is to alert us to the question of what method of calculation was being employed in the measurement of distances or sizes of heavenly bodies: (i) calculations that were to some extent scientific, (ii) a Pythagorean notion of notes on a musical scale, and (iii) simple non-musical numerical proportions.

35 Sambursky, 1956, p. 15.

36 Cf. Kirk-Raven, 1957, p. 134; Burkert, 1972, p. 417; Jaeger, 1939, p.137.

37 The names of the architects identified with the archaic Heraion, and the archaic Artemision, come to us from Vitruvius, c. 25 B.C. in his *Ten Books on Architecture.* He mentions that they wrote prose treatises, 7.12, p. 198. That these architects wrote prose treatises, no longer extant, is doubted by some, but important scholars have accepted Vitruvius' assertion. If they did write prose treatises at roughly the same time that Anaximander wrote his philosophical book in prose, another possible connection between their communities would be suggested. Cf. Coulton, 1977, p. 24: "An important development in the middle of the sixth century was the writing of the first architectural treatises. . . . These must have been among the earliest prose works in Greek, for the first philosophical work in prose was written by Anaximandros of Miletus at just about the same time. The Ionian school of philosophy in the sixth century had an interest in the practical as well as the abstract. . . . It is presumably not merely coincidental, therefore, that the first Greeks to write about architecture were working in Ionian cities." Cf. also Hurwit, "Rhoikos and Theodorus wrote a book about their limestone behemoth – another example of early prose and one probably far more prosaic than Anaximander's book on nature . . . [sc. concerning the Artemision]. Chersiphron of Knossos and his son Metagenes wrote a book about their temple, too." Cf. also Dinsmoor, 1902/1950, p. 124n1: "The book by Theodorus (the earliest architectural treatise of which the title has come down to us) . . . ." Cf. also Tomlinson, 1976, p. 127: "The architects of the temple [of Hera] were Rhoikos and Theodorus. Theodorus wrote a treatise about it, which was known to the Roman architect Vitruvius."

38 Cf. Herodotus, 2.152-4, and also 4.152.1.

39 Boardman, 1980, pp. 110-115; Lawrence, 1962, pp. 132-133; Berve and Gruben, 1960, pp. 445, 447, 454; Dinsmoor, 1902/1950, pp. 124-125; Robertson, 1929/1983, p. 60; Coulton, 1977, pp. 32-38. Tomlinson, 1976, p. 125; Grant, 1987, p. 153; Braun, 1982, pp. 32-56.

40 Clarke and Engelbach, 1930/1990, pp. 46-68.

[41] Ibid. p. 47. Cf. also Coulton, 1977, p. 52: "The idea of an architectural ground plan had certainly been developed in both Egypt and Mesopotamia, for examples have survived." In this context, he also mentions the statue of Gudea of Lagash (c. 2200 B.C.) showing him seated with a drawing table on his knees, equipped with a stylus and ruler; on one of these tables a plan is engraved.

[42] Ibid. p. 49.

[43] Ibid. p. 51.

[44] Ibid. pp. 52-53.

[45] Kemp, 1991, p. 139.

[46] Ibid. p. 138, where he speculates about the building practices of the Egyptian architects. He places much greater emphasis on planning and execution of the construction on-site rather than by mathematical plans.

[47] Haselberger, 1985, pp. 126-129.

[48] Ibid. p. 132.

[49] Kienast, 1985, p. 111, n.21: "Die Diskussion, ob der griechische Architekt seinen Entwurf nur in schriftlicher Form oder auch durch Zeichnungen festlegte, ist mittlerweile durch die bedeutende Entdeckung in Didyma um einen wichtigen Schritt weitergekommen: Es gab zumindest auch - maßstabsgerechte - Zeichnungen. s. L. Haselberger, Ist Mitt 30, 1980, 191ff. und ders., *Architectura* 13, 1983, 13."

[50] Cf. Coulton, 1977, p. 53; he argues that evidence found in the fifth and fourth centuries "must be applicable, if in a simpler form, to the sixth century." But, at the time of his writing, Haselberger's evidence was unknown.

[51] Cf. Coulton, 1977, p. 38, the example from Perachora (c. late 8th century); also the house models in the Samos museum in Vathi, from the 7th century.

[52] Coulton, 1977, p. 53. Note: as recently as 1977, this was a broadly accepted opinion.

[53] Kienast, 1985, pp. 111: "Die Aufschnürung, jener entscheidende Vorgang, bei dem der Architekt seinen Entwurf am Bauplatz in wirkliche Maße überträgt, läßt sich bei fast allen griechischen Bauten nachweisen. Sichtbare Zeugen dieser Aufschnürung sind in der Regel kurze Anritzungen von bestimmten Achsen und Ecken, die die Gestalt des Baus charakterisieren. Im Gegensatz dazu handelt es sich beim Schatzhaus D um eine vollständige Grundrißzeichnung."

[54] Kienast, 1986/87; cf. also the earlier and much shorter piece, 1977, pp. 97-116; cf. also Felsch and Kienast, 1973 and 1975.

[55] Ibid. pp. 232-237. This idea that Eupalinos perhaps invented his own tunnel measure since the increment of measure is not in Samian ells, nor

in any unit of construction known on the mainland, is interesting. In a not unrelated matter, Dilke, 1987, p. 13, and 1985, p. 81, had suggested that if Anaximander made a terrestrial map that included marked out distances, he would have needed some form of numeration, in abbreviated notation; Dilke then wondered if the Milesian form of numeration might not be traceable back to Anaximander. The architects and philosophers were both involved in activities that led to inventing their own measures?

56 According to a recent, and yet unpublished excavation by Kienast that revealed pottery just under the Rhoikan foundation dating to 575 B.C., the dating of the beginning of that construction can be reliably fixed. In addition, for a possible connection of "Rhoikos" with Egypt, cf. Boardman, 1980, p. 132, who wonders if a multiple eye cup dedicated by someone named Rhoikos to Aphrodite at Naucratis (c. 575-550 B.C.) might not be by the architect of the archaic Heraion. This consideration arises in the context of discussing Egyptian influence in Ionic temple building. It is also noteworthy, with regard to the Egyptian connection, that Pliny, *Natural History*, xxxvi, 90, refers to the Samian Heraion, identified with Rhoikos and Theodorus, as "The Labyrinth" indicating that the inspiration was probably the Egyptian temple by Lake Moeris referred to by Herodotus (II. 148) under that name. For the historical background in Samos, cf. Shipley, 1987.

57 There is considerable disagreement over the exact measurements, or at least the way those measurements should be presented. Some have given the dimensions in terms of the rectangle formed by the stylobate while others have given them for the larger rectangle constituted by the inclusion of the two steps: Kyrieleis, 1981, p. 73 (and 1980, pp. 336-350) following Walter, 1976, gives the measurements 172.2 by 344.4 feet (52.5m x 105m = 100 x 200 Samian ells); Dinsmoor, 1902/1950, p. 124, and Tomlinson, 1976, both give the same measurements: 174 by 314 feet, or 171 by 311 feet depending upon the reference to steps. Robertson, 1929/1983, p. 331: 50.50m x 103m. The possible elevation reconstruction follows Tomlinson, 1976, p. 125.

58 There is, again, disagreement over the precise measurement, perhaps as a result of measuring from the lowest stair or restricting the dimensions to the stylobate. Bammer, 1984, p. 183, and Akurgal, 1985, p. 148: 180.9 by 377.4 feet (55.10m x 115m); Dinsmoor, 1902/1950, in the Appendix, and Tomlinson, 1976, p. 129: 55.1m x 115.14m; Robertson, 1929/1983, p. 331: 55.10m x 109.20m. The possible elevation reconstruction follows Grant, 1987, the illustration after p. 204. For the measurements and dating of the archiac Didymaion, cf. Gruben, 1963.

59 Coulton, 1977, p. 45.

60 Ibid. p. 31.

61 However, cf. the interesting work by Peronotis, 1972, who argues for the use of architectural drawings.

62 Ibid. Cf. p. 51, from which I have drawn on considerably in phrasing this paragraph.

63 The assignment of the date 548 B.C. is not controversial since many commentators accept c. 550 B.C. Nor is the argument for establishing the date an issue of contention. According to the tradition traced through Apollodorus (cf. Jacoby, 1902, pp. 210ff; also Kirk-Raven, 1971, pp. 101-102), Anaximander's book appeared one year before the conquest of Sardis by Cyrus. Anaximander's age is known not by his floruit and not by his death (although close to it); it is established by something in his book, a book not identified with his flourishing at forty but with the publication of his thoughts preciously close to the end of his life. Burnet, 1945, p. 13, inferred from Diogenes' testimony that the chronographer Apollodorus found definite evidence, perhaps in a summary version of his book, that Anaximander was sixty-four in 547/6. Concerning "publication," I follow Burkert, 1985, p. 310; Heraclitus' dedication of his book in the temple of Artemis (cf. Diogenes Laertius, 9.6) was the act of making the book public, that is, publishing it. Whether Anaximander dedicated his book at the temple of Apollo in Didyma, or elsewhere, we cannot say, but this is how I make sense of a "publication" in the archaic period.

64 Cf. Orlandos, 1965, II., pp. 100ff; Martin, 1965, pp. 226ff; Berve and Gruben, 1960, pp. 444-467; Lawrence, 1962, pp. 132-133. Compare to Clarke and Engelbach, 1930/1990, pp. 136-150.

65 Cf. Vernant, 1965/1983, p. 283. Aristotle, in *Metaphysics* A, claims that philosophy begins with the experience of wonder, *thaumazein*.

66 DK 12A6.

67 DK 12A6.

68 Cf. the discussion in Dilke, 1985, pp. 22-23, and 56. These maps were either painted on wood or worked in bronze, like the bronze tablet that Aristagoras brought to Sparta, according to Herodotus (5.49), in order to win assistance for the Ionian revolt. No early Greek maps survive, but there is in the British Museum a clay tablet belonging to the neo-Babylonian or Persian date (roughly 600 B.C.). Cf. Kahn, 1960, who reprints the map (Plate 1).

69 DK 12A1. Diogenes Laertius, II, 1-2. Cf. also Gibbs, 1976, pp. 2-3, and her reflections on the report of Hunt, 1946, in suggesting a possible reconstruction for Anaximander's sundial. Cf. also Szabo, 1977, pp. 341-357; Sarton, 1952, vol. I, p. 175.

70 Heidel, 1937, pp. 7-17. Cf. also the discussion of this point in H. D. P. Lee's commentary, p. 103, to Aristotle's *Meteorologica*. Loeb Series.

71 Robinson, 1968, p. 19. Cf. also the rendition by Brumbaugh, 1964, p. 22.

72 Some might prefer to call the illustration a "horizontal cross-section" rather than a "plan." This is because the term "plan" tends to connote absolute directions, let's say, of up and down, top and bottom. If one accepts the testimony, derived from Hippolytus and Aetius – cf. Kahn, 1960, p. 56, and 84-85 – that Anaximander claimed the existence of antipodes, creatures who lived on the other side of the earth (i.e., the horizontal surface parallel to the one on which we live), then the idea of absolute directions, up and down, left and right is discredited. Vernant, 1983, pp. 179ff, holds just this position on the issue that for Anaximander absolute value is no longer attached to directions in space as it was in Hesiod and others. Thus, if Anaximander abandons a view of absolute spatial relations, then "plan" may prove to be misleading. However, a horizontal cross-section of the cosmos through the earth, from either our point of view or that of the antipodes, will produce the same picture.

73 Cf. Lawrence, 1962, pp. 225ff; Orlandos, 1965, II, p. 100ff; Martin, 1965, pp. 193-199.

74 Martin, 1965, pp. 195-196.

75 For the Egyptian technique, cf. Clarke and Engelbach, 1930/1990, pp. 99-109. Lawrence, 1962, p. 225, claims that the *anathyrosis* technique originates in Egypt. Coulton, 1977, denies the technique to Egyptian architecture, p. 47, but then modifies his position to note, p. 169, n. 73, that Egyptian masonry does present vertical joints prepared in this fashion but only on the outer face. His point is that since the blocks do not have their rear faces dressed, they do not exhibit true *anathyrosis*. For the argument here, it is sufficient to observe that the *anathyrosis* technique in some form is displayed in the Egyptian masonry that the Ionian Greeks could have observed.

76 Coulton, 1977, pp. 46-47.

77 Cf. Nylander, 1962, p. 47, figs. 56-60.

78 Cf. Orlandos, 1965, II, pp. 100-101.

79 Cf. Diels, 1897, pp. 228-237 (esp. 231) for the earliest diagram I have been able to find.

80 Heath, 1913, pp. 35-36; he refines the diagram offered by Neuhauser, 1883, pp. 427-428. The only other Anaximander diagram in an English language publication that I know of is in Rescher, 1958, pp. 718-731.

81 Couprie, 1989, p. 227-231.

82 However, the account of meteorological phenomena offers us some reason to suppose that changes in the sun's altitude, for example, may be due to the winds. Concerning mechanical explanation in Anaximander's meteorology, cf. Kahn, 1960, pp. 98-100, and more general discussions on

mechanism in nature, cf. Heidel, 1909/1910, pp. 77-113; ; Reinhardt, 1926; pp. 161-176; pp. 387-395; Rescher, 1958, pp. 718-731; de Solla Price, 1974.

83 Furley, 1987, pp. 23-27; 1989, pp. 14-22.

84 Aristotle, *De caelo*, II.13 (295b10ff): "The majority of thinkers, then, debate over these causes [mentioned above]. But some say that it is because of 'equilibrium' (*homoiotēta*) that the earth remains at rest, as among the ancients, Anaximander. For that which is situated in the middle and is equally related to the extremes, is not obliged to move in one direction rather than another, either up or down, or sideways; and because it is impossible to move simultaneously in opposite directions, it necessarily remains at rest." For the translation of *homoiotēs*, various renditions have been adopted: "Similarly": Kahn, 1960, pp. 76, 79n3; Lloyd, 1978, p. 68; "Indifference": Guthrie, 1962, p. 98; Furley, 1989, p. 16; Robinson, 1972, p. 111, and 117n1; "Equilibrium": Vlastos, 1953/1970, p. 75; Kirk-Raven, 1957, p. 134; "Equal Distance": Cornford, 1952, p. 165; "Equiformity": Dicks, 1970, p. 44. Other renditions proposed include "likeness" and "uniformity." Despite the variations in translation, the meaning does not seem to be in doubt.

85 Heidel, 1906, pp. 279-282; and 1937, pp. 68-69.

86 Robinson, 1971, pp. 111-118, first presented to a meeting of the SAGP in 1953.

87 Cf. Furley, 1987, p. 26, and 1989, p. 22; cf. also Robinson, 1971, p. 116. The passage from Simplicius' commentary on *De caelo* is 532.13.

88 At an early stage of reflecting upon the organization and patterns exhibited by the temple columns, I focused exclusively on the number of columns and the patterns established by them. Only afterwards did it seem that this approach was leading nowhere. I then turned to focus on the number and variety of spaces rather than the columns. The spaces, opened and closed by the arrangement of the columns, seem to orchestrate the movement and feeling generated by the building. The symphony of feeling seemed to be a product of limiting and unlimiting the spaces.

89 Mark Johnson, 1987, and 1992.

90 Mark Rollins, 1989.

91 Roger Shepard, 1984.

92 These projects follow from Kuhn's lead, 1962/1970, although not always in the particular ways he anticipated. The emphasis must be placed on the importance of case studies and away from broad and sweeping generalizations that characterized the positivist approach.

BIBLIOGRAPHY

Akurgal, Ekrem. *Ancient Civilizations and Ruins of Turkey.* Istanbul: Haset Kitabevi, 1985.

Anton, John P., and George L. Kustas (eds.). *Essays in Ancient Greek Philosophy*. Albany: State University of New York Press, 1971.

Bammer, Anton. *Das Heiligtum der Artemis von Ephesos*. Graz: Akademische Druck - u. Verlagsanstalt, 1984.

Barnes, Jonathan. *The Presocratic Philosophers*. 2 vols. London: Routledge and Kegan Paul, 1979; 2nd ed. 1982.

Berve, H., and G. Gruben, *Greek Temples and Theatres and Shrines*. New York: Harry N. Abrams, 1960.

Boardman, J. *The Greeks Overseas: Their Early Colonies and Trade*, rev. ed. London: Thames and Hudson, 1980.

Braun, T. F. R. G. "The Greeks in Egypt," in *Cambridge Ancient History*. 2d edition, III.3. Cambridge: Cambridge University Press, 1982, pp. 32-56.

Brumbaugh, Robert S. *The Philosophers of Greece*. New York: Thomas Crowell, 1964.

Burkert, Walter. *Lore and Science in Ancient Pythagoreanism*. Cambridge, Mass.: Harvard University Press, 1972. Trans. E. L. Minar, Jr., with revisions from *Weisheit und Wissenschaft: Studien zu Pythagoras, Philolaos und Platon*. Nuremberg: Verlag Hans Carl, 1962.

Burkert, Walter. *Homo Necans*. Berkeley and Los Angeles: University of California Press, 1983. Trans. Peter Bing from the German *Homo Necans*, 1972.

Burkert, Walter. "Iranisches bei Anaximandros." *Rheinisches Museum*, 106 (1963), pp. 97-134.

Burkert, Walter. *Greek Religion*. Trans. J. Raffan. Cambridge, Mass: Harvard University Press, 1985.

Burnet, John. *Early Greek Philosophy*. 1st ed., 1892; 4th ed. repr. London: Adam and Charles Black, 1945.

Clarke, S., and R. Engelbach. *Ancient Egyptian Masonry*. New York: Dover Press, [1930] 1990.

Cornford, Francis M. *Principium Sapientiae: The Origins of Greek Philosophical Thought*. Cambridge: Cambridge University Press, 1952.

Coulton, J. J. *Greek Architects at Work*. Ithaca: Cornell University Press, 1977.

Couprie, D. L. *De verordening van de Tijd: interpretatie en vertaling van het fragment van Anaximander met een appendix over de visualisering van zijn wereldbeeld*. Academisch Proefschrift. Filosofische Reeks No. 30, 1989.

Detienne, M. *Les Maitres de vérité dans la grèce archaique*. Paris, 1967.

Dicks, D. R. "Solstices, Equinoxes, and the Presocratics." *Journal of Hellenic Studies* 86 (1966), pp. 26-40.

Dicks, D. R. *Early Greek Astronomy to Aristotle.* London: Thames and Hudson, 1970.

Diels, Hermann. "Ueber Anaximanders Kosmos." *Archiv für Geschichte der Philosophie* 10 (1897) pp. 228-237.

Diels, Hermann. *Doxographi Graeci.* Berlin: de Gruyter, 1879; repr., 1958.

Diels, Hermann, Kranz, Walther. *Die Fragmente der Vorsokratiker* (DK). 6th ed. Berlin: Weidmann, 1951-2.

Dilke, O. A. W. *Greek and Roman Maps.* London: Thames and Hudson, 1985.

Dilke, O. A. W. *Mathematics and Measurement.* London: British Museum Publications, 1987.

Dinsmoor, W. B. *The Architecture of Ancient Greece,* 3d ed. (first published in 1902). New York: Norton, 1950.

Diogenes Laertius. *Lives of the Philosophers.* 2 vols. Trans. R. D. Hicks. Loeb Series. Cambridge, Mass: Harvard University Press, 1966.

Farrington, B. *Greek Science.* Part I, London: Penguin Books, 1944; new ed. 1949; Part II, 1949; first publication as one vol. 1953; rev. ed., 1961.

Felsch, R. C. S., and Kienast, H. "Die Wasserleitung des Eupalinos." *Archäologischer Anzeiger* (1973), pp. 401-14.

Felsch, R. C. S., and Kienast, H. J. "Die Wasserleitung des Eupalinos." *Archäologischer Anzeiger* (1975), pp. 19-35.

Frischer, B. *The Sculpted Word.* Berkeley and Los Angeles: University of California Press, 1982.

Furley, David J. *The Greek Cosmologists,* vol 1. Cambridge: Cambridge University Press, 1987.

Furley, David J. *Cosmic Problems.* Cambridge: Cambridge University Press, 1989.

Gernet, Louis. *The Anthropology of Ancient Greece.* Baltimore: The Johns Hopkins University Press, 1981 (first published 1917).

Gibbs, Sharon L. *Greek and Roman Sundials.* New Haven, Conn., and London, 1976.

Goody, J., and Watt, I. P., "The Consequences of Literacy." *Literacy in Traditional Societies.* Cambridge: Cambridge University Press, 1968, pp. 27-68.

Grant, Michael. *The Rise of the Greeks.* New York: Macmillan, 1987.

Gruben, G. "Das archaische Didymaion." *Jahrbuch des deutschen archäologischen Institut,* 78 (1963), pp. 78-177.

Guthrie, W. K. C. *Aristotle: On the Heavens.* Loeb Classical Library. London: Heinemann, 1939; repr., 1953.

Guthrie, W. K. C. *A History of Greek Philosophy,* vol. 1. Cambridge: Cambridge University Press, 1962.

Hahn, Robert. "What Did Thales Want to Be When He Grew Up? or, Re-Appraising the Roles of Engineering and Technology in the Origins of Early Greek Philosophy/Science," in *Plato, Time, and Education: Essays in Honor of Robert S. Brumbaugh,* ed. Brian Hendley. New York: State University of New York Press, Albany, 1987.

Haselberger, Lothar. "The Construction Plans for the Temple of Apollo at Didyma." *Scientific American*, vol. 253, no. 6, December 1985, pp. 126-132.

Heath, Sir Thomas. *Aristarchus of Samos: The Ancient Copernicus.* Oxford, 1913; repr., Oxford: The Clarendon Press, 1959.

Heidel, W. A. "The *DINE* in Anaximenes and Anaximander." *Classical Philosophy* 1 (1906), pp. 279-82.

Heidel, W. A. "*Peri Phuseos*: A Study of the Conception of Nature among the Pre-Socratics." *Proceedings of the American Academy of Arts and Sciences,* 45 (1909-1910), pp. 77-113.

Heidel, W. A. *The Frame of the Ancient Greek Maps, with a Discussion of the Discovery of the Sphericity of the Earth.* American Geographical Society Research Series, 20. New York: American Geographical Society, 1937.

Hirzel, R. *Themis, Dike, und Verwandtes.* Leipzig, 1907.

Holloway, R. R. "Architect and Engineer in Archaic Greece." *Harvard Studies in Classical Philosophy,* 73 (1969), pp. 281-290.

Hunt, D. W. S. "An Archaeological Survey of the Classical Antiquities of the Island of Chios Carried Out Between the Months of March and July, 1938." *The Annual of the British School at Athens,* no. 41, session 1940-1945, pp. 41-42.

Hurwit, J. M. *The Art and Culture of Early Greece 1100-480 BC.* Ithaca: Cornell University Press, 1985.

Jacoby, F. *Die Fragmente der griechischen Historiker.* Berlin 1923-30, and Leiden 1940-58, E. J. Brill.

Jaeger, Werner. *Paideia: The Ideals of Greek Culture.* Trans. G. Highet. 3d ed. Oxford: Blackwell, 1945.

Johnson, Mark. *The Body in the Mind: The Bodily Basis of Meaning, Imagination, and Reason.* Chicago: University of Chicago Press, 1987.

Johnson, Mark. *Moral Imagination.* Chicago: University of Chicago Press, 1992.

Kahn, Charles H. *Anaximander and the Origins of Greek Cosmology.* New York: Columbia University Press, 1960.

Kahn, Charles H. "On Early Greek Astronomy." *Journal of Hellenic Studies,* 90 (1970), pp. 99-116.

Kemp, Barry J., *Ancient Egypt: Anatomy of a Civilization.* London and New York: Routledge, 1989.

Kienast, H. J. "Der Sog. Temple D im Heraion von Samos. Part I: Ein Schatzhaus aus der Nachpolykratischen Zeit." *Mitteilungen des Deutschen Archäologischen Instituts. Athenische Abteilung.* Berlin: Gebr. Mann Verlag. Band 100, 1985, pp. 105-127.

Kienast, H. J. "Der Tunnel des Eupalinos auf Samos." *Architectura* (1977), pp. 97-116.

Kienast, H. J. "Der Tunnel des Eupalinos auf Samos," in *Mannheimer Forum*, 86/87, pp. 179-241.

Kirk, G. S. and Raven, J. E. *The Presocratic Philosophers: A Critical History with a Selection of Texts.* 1957. 2d ed. by G. S. Kirk, J. E. Raven, and Malcolm Schofield. Cambridge: Cambridge University Press, 1983.

Kuhn, Thomas S. *The Structure of Scientific Revolutions.* Chicago: University of Chicago Press, 1962. 2d ed., 1970.

Kyreileis, H. "Ausgrabungen im Heraion von Samos 1979." *Archäologischer Anzeiger* (1980), pp. 336-50.

Kyreileis, H. *Führer durch das Heraion von Samos.* Athens: DAI/Ekdotike Ellados, 1981.

Lawrence, A. W. *Greek Architecture.* 2d ed. Baltimore, Md.: Penguin Books, 1962.

Lloyd, G. E. R. *Magic, Reason, and Experience: Studies in the Origin and Development of Greek Science.* Cambridge: Cambridge University Press, 1979.

McKirahan, Richard. *Philosophy before Socrates.* Bloomington: Indiana University Press, 1994.

Neuhauser, I. *Anaximander Milesius sive vetustissima quaedam rerum universitatis conceptio restituta.* 1883.

Nylander, C. *Opuscula Atheniensia,* 4 (1962) , 47, figs. 56-60.

O'Brien, D. O. "Anaximander's Measurements." *The Classical Quarterly, 17* (1967), pp. 423-432.

Orlandos, A. *Les Materiaux de construction: Et la Technique architecturale des anciens Grecs.* 2 vols. Trans. from the modern Greek by V. Hadjimichali. Paris: Editions E. De Boccard, 1966, 1968.

Pliny the Elder, *Natural History.* Loeb Series. Cambridge, Mass: Harvard University Press.

Price, D. J. de Solla. *Gears from the Greeks: The Antikythera Mechanism. Transactions of the American Philosophical Society,* n.s. 64, pt. 7, 1974.

Reinhardt, Karl. *Kosmos und Sympathie.* Munich: Beck, 1926.

Rescher, Nicholas. "Cosmic Evolution in Anaximander." *Studium Generale,* 11 (1958), 718-31. Repr. in Rescher, *Essays in Philosophical Analysis.* Pittsburgh: University of Pittsburgh Press, 1969, pp. 3-32.

Robertson, D. S. *Greek and Roman Architecture.* Cambridge: Cambridge University Press, [1929] 1983.

Robinson, John Mansley. *An Introduction to Early Greek Philosophy: The Chief Fragments and Ancient Testimony, with Connecting Commentary.* Boston: Houghton Mifflin, 1968.

Robinson, John Mansley. "Anaximander and the Problem of the Earth's Immobility," in Anton and Kustas, *Essays in Ancient Philosophy,* vol. I. Albany: State University of New York Press, 1971, pp. 111-18.

Rollins, Mark. *Mental Imagery.* New Haven: Yale University Press, 1989.

Sambursky, S. *The Physical World of the Greeks.* 3 vols. Trans. from the Hebrew by Merton Dagut. Princeton: Princeton University Press, 1956.

Sarton, George. *A History of Science,* vol. 1. London and Cambridge, Mass: Harvard University Press, 1952.

Shepard, Roger. *Mental Images and their Transformations.* Cambridge: MIT Press, 1982.

Shipley, Graham. *A History of Samos 800-188 B. C.* Oxford: The Clarendon Press, 1987.

Snodgrass, A. M. *Archaic Greece: The Age of Experiment.* Berkeley and Los Angeles: University of California Press, 1980.

Szabo, Arpad. "Anaximandros und der Gnomon." *Acta Antiqua,* 25 (1977), pp. 341-357.

Tomlinson, R. A. *Greek Sanctuaries.* London: Paul Elek, 1976.

Vernant, J. P. *Les Origines de la pensèe greque.* 1962. Trans. as *The Origins of Greek Thought.* Ithaca: Cornell University Press, 1982.

Vernant, J. P. *Mythe et pensèe chez les Grecs.* 1965. Trans. as *Myth and Thought among the Greeks.* London: Routledge and Kegan Paul, 1983.

Vidal-Naquet, Pierre. *The Black Hunter.* Trans. A. Szegedy-Maszak. Baltimore and London: The Johns Hopkins University Press, 1986.

Vitruvius. *The Ten Books on Architecture.* Trans. M. H. Morgan. New York: Dover, 1968.

Vlastos, Gregory. "Equality and Justice in Early Greek Cosmologies." *Classical Philology* 42, (1947), pp. 156-178.

Vlastos, Gregory. "Isonomia." *American Journal of Philology,* 74:4 (1953), pp. 337-366.

Vlastos, Gregory. *Plato's Universe.* Oxford: Oxford University Press, 1975.

Walter, Hans. *Das Heraion von Samon. Ursprung und Wandel eines griechischen Heiligtums.* Munich: R. Piper, 1976.

West, M. L. *Early Greek Philosophy and the Orient.* Oxford: The Clarendon Press, 1971.

PAUL B. THOMPSON

TECHNOLOGICAL VALUES IN THE APPLIED SCIENCE LABORATORY[1]

The goal of this paper is to frame a philosophical question that has received very little attention from anyone, and almost none from professional philosophers. While "social constructionism" has introduced us to the view that the various theories, schools, and paradigms of science are as much the result of social forces as they are the realization of nomothetic explanatory ideals, the new "social" approaches to the philosophy of science have failed to take up obvious normative issues associated with the use of science to accomplish social objectives. Applied science wears its allegiance to the goodness of curing disease, building bridges or producing food on its sleeve, yet studies that have raised questions about the social impact of medical, engineering, or agricultural technologies have, for the most part, been reluctant to press those questions at the level of choosing to conduct research that was intended to result precisely in the technologies and techniques that are questioned. If we expect to impede or direct technological change, it is immanently reasonable to think that the most effective point of intervention will be at the stage of research choice.

Such an intervention need not be anti-technology. Indeed, the most likely form that influence on research choice is likely to take is one of directing researchers to produce technology, though perhaps technology of a different sort than is produced now. There have been several moments in the past two decades when detectable amounts of political enthusiasm for research on "alternative" technological goals have emerged. There have been calls for research on alternative fuels and on recycling, for example. It is not obvious, however, how such calls for applied research are to be justified philosophically, nor is it obvious that scientists have a responsibility to respond. The agenda for research in applied areas is a complex result of public pressure, social norms, and the structure of scientific institutions.

Joseph C. Pitt (ed.), New Directions in the Philosophy of Technology, 139-151.

There are several philosophical questions to be posed with respect to the research agenda, but this paper is pointed at one which reaches into the institutional structure of applied science. Although many applied scientists work in private development groups dedicated to the production of marketable technologies, many work in non-profit laboratories and universities which are organized around disciplinary departments such as physiology, entomology, or electrical engineering. Virtually all applied scientists are trained in such a structure. Is it obvious that disciplinary organization in the applied sciences serves the explicit social goals that applied science is expected to achieve? Does the disciplinary structure of applied science itself impose directionality on technical change and, hence, upon fundamental political issues? If so, what, if anything, should be done about it? These are the questions that this paper (imperfectly, I fear) attempts to raise. The goal is to give the questions a philosophical shape and to prompt a wider reflection upon them. Answers will not be forthcoming.

1. THE POLITICAL ECONOMY OF RESEARCH CHOICE

In 1983, Lawrence Busch and William Lacy published the results of their sociological study of research choice in agricultural science. Busch and Lacy were attempting to show how granting agencies, tenure and promotion and peer reviews, and influence exerted through state legislators, farm organizations, and agribusiness shaped the agenda for research in agriculture. Their aim was to develop a "political economy" approach to the sociology of science that would contrast with Mertonian models that stressed organization of scientific disciplines around shared theoretical assumptions, research methods and discovery goals (Busch and Lacy, 1983). While there has been a significant appreciation of the philosophical implications of the Busch and Lacy study among those interested in agriculture, it has not been widely related to the philosophy of technology. Some philosophers of technology, following Heidegger, have stressed the ontological or historical priority of technology over science; others have analyzed the way in which existing technology imposes an institutional framework upon the activity of scientists. Both of these themes share Busch and Lacy's rejection of science as a pattern of inquiry guided autonomously by scientists in pursuit of rigor, parsimony and truth. Nevertheless, by focusing on an applied science devoted to the solution of practical problems, Busch and Lacy's work demands that we examine a link between philosophy of science and philosophy of technology that is both more obvious and more subtle than these.

In being configured to solve problems in manufacturing, in agriculture, or in medicine, applied sciences are, by definition, intended to produce technologies such as materials, machines, plant varieties, drugs, and surgical techniques. There is no possibility of even a pretense that the organization and conduct of applied research is *purely* an attempt to describe or explain the world accurately. Applied sciences are geared toward the development of technology from the outset, at least in so far as technology is understood as knowing "how to do it," whatever "it" is. The political economy approach to the sociology of science suggests that the choice of which technologies to develop is determined by "material" forces such as the granting agencies and interest groups already mentioned. As such, it opens a direct route to philosophical questions about distributive justice, social efficiency, human rights and violations of liberty as they relate to scientific research. If research choice is determined by "the forces of production," it is important to raise ethical questions about how applied science serves to concentrate and distribute social power. The political economy approach shows why the research agenda is simultaneously a technological choice and a political choice, at least for the applied sciences.

The political economy approach also suggests that change in research directions can be precipitated simply by a realignment of the rewards system for applied scientists. If applied science follows money (to oversimplify), the direction of research should be unilaterally correlated with availability of funds. By contrast, I will argue that technological values operating within applied science laboratories constrain the influence of the rewards system upon problem choice. While I think that what I will say is true for applied science in general, the main impetus for my reflections is the current state of agricultural research. Over two decades, agricultural research has been subjected to repeated criticism for emphasizing productivity and yield enhancing technology at the expense of environmental quality, small farm and rural community development, human health and safety, and just, sustainable development in the Third World. While there was some initial resistance to these criticisms, they have now been very substantially accepted by scientists and administrators within most major research organizations and funding agencies. During the past decade the alignment of political interest groups has also shifted to substantially favor alternative streams of agricultural research. Streams that are applied not to the enhancement of productivity, but to the goals advocated by the critics of the 1970s. Despite these shifts, the evidence is that even those research projects nominally dedicated to low input or sustainable alternatives continue to be dominated by the productivity paradigm.

2. PRODUCTIVITY AS VALUE IN AGRICULTURAL RESEARCH

The agricultural scientist's emphasis upon productivity is ethically defensible on *prima facie* grounds because increases in food availability obviously serve the comprehensive ethical norms of social benefit. Productivity goals help shape a researcher's choice of topics in much the same way that the goal of ending disease shapes the medical researcher's choice of topics. It is, in other words, a norm that provides a certain structure to the agricultural researcher's activity, and provides a rolling horizon for measuring the success or failure of any given research stream. It is quite common for language expressing an experiment's or research finding's contribution to agricultural productivity to be included in grant proposals, in tenure and promotion reviews, in awards recognizing the achievements of a particular researcher, as well as in technical publications themselves. In what follows I shall refer to this norm simply as "productivity," though it is clear that what is meant is a collective value judgment regarding diverse ways of manipulating organisms to increase yields, to decrease losses in yield and to increase production revenue relative to input costs.

Despite its broad justifiability, productivity can be qualified by other values. In recent decades, agricultural scientists' seemingly unilateral emphasis upon productivity has been criticized on at least three broad fronts. One is environmental quality. Agricultural techniques that increase productivity may have unintended harmful impacts upon environmental quality. A second is international justice. The ethical rationale for increasing productivity is that it advances the well being of the majority, but some have questioned whether this rationale holds true in international assistance contexts. The third criticism originates from an alleged trade-off between increases in productivity and a decline in the viability of small to medium sized family farms in the United States. Here, critics argue that preservation of family farms is more important than increasing net farm productivity. Essays collected by Dahlberg (1986) and by Thompson and Stout (1991) illustrate this point.

I shall not assess the merits of these criticisms in this paper. I mention them because I want to point out the dialectical opposition that arises in each case. Productivity is seen as a valid but problematic thesis guiding agricultural research. In each case, an antithetical value judgment opposes the value of productivity, and by implication, agricultural research as such. Agricultural researchers are faced with a choice. Either they reject the arguments of the critics, and defend productivity against the values that stand in contradiction to it, or they must find a synthesis, some way of accommodating the opposing values within the framework of agricultural research. The first of these alternatives may be the one that has been chosen most often. It requires a defense of productivity that either denies the alleged link between agricultural technology and its unwanted consequences, or defends productivity and economic growth in the conventional terms of political philosophy. It is the possibility of opting for synthesis, however, that is of interest here, for this alternative requires scientists to internalize new goals, or, at least, to qualify research-defining values in ways that change the very concept of their disciplinary activity.

The problem now seems to be that while the scientific staffs of agricultural research organizations have the capacity to do work that addresses productivity, they lack the capacity to propose or conduct research that addresses environmental quality and other goals. Scientific administration is largely a task of coordinating individual research efforts, so the capacities of existing staff place severe short term constraints upon choice of research problems. How can this lack of capacity be understood as a defect of philosophical values?

3. REDUCTIONISM

Critics of productivity have traced constraints on research choices to what they call reductionism in agricultural science. Miguel Altieri (1987), and Baird Callicott (1990), for example, have argued that mechanistic reductionism, or "physics envy," in biology represents an explanatory program incapable of recognizing (much less modeling) systems level interactions crucial for an ecological understanding of agriculture. Billie DeWalt (1988) has made a similar criticism of reductionist paradigms in agricultural science, but uses the argument to account for inattention to social and cultural dimensions of agricultural production. To be sure, undue emulation of physics may occur in applied biology, but it is a *non sequitur* to attribute the neglect of alternative research priorities to this form of reductionism. In the present context, reductionism must be interpreted to mean the strategy of breaking broad research goals (such as productivity or environmental quality) into discrete, individualized research efforts that utilize a specific research capacity tied to a discipline, a lab, or, perhaps, to an individual scientist. While this idea was clearly a part of Descartes' method for the sciences, it is far from the emphasis upon reducibility to physics that is sometimes the focus of the reductionism debate in philosophy of science.

It is useful to think of applied science reductionism as a function of the research capacity implicit in the existence of a laboratory. The laboratory includes the equipment, materials and personnel required to conduct experiments of a certain type. The idea of the laboratory can be used to refer generally to the relevant research unit, whether it be a single scientist at a desk, or a research team spread across several locations. There is certainly a range of experiments that can be conducted in any given lab, but the tendency of modern science is for this range to be comparatively narrow. Range is restricted by material equipment and by the technical and disciplinary expertise of the scientific staff. Technology and expertise are also what gives the lab its capacity, so we should expect some trade-off between capacity and range (Latour, 1988). A major component of capacity resides in the laboratory's power to control events within its confines. Both technology and expertise help control extrinsic events, so that only correlations of specific interest can account for what happens in the lab.

The applicability of agricultural research depends partially upon a farmer's ability to replicate, or at least approximate, some of the controlled conditions that exist in a lab. Although the open fields which serve as labs for some forms of agricultural research are highly uncontrolled when compared to research in the basic sciences, they are far more controlled than a real farmer's fields. Social variables and human factors, in particular, are highly controlled, and environmental variables, though not controlled in a strict sense, may be very unrepresentative of the farmer's field. Even so, the control that exists in an agricultural scientific laboratory may be quite conducive to certain kinds of productivity enhancing research. To the extent that increases in yield can be defined in physiological terms, for example, the laboratory can control extraneous factors in a search for the biological factors that correlate to increased yield. These results can be transferred to the farmer's field by transforming the factors that give the scientist control over the lab into factors that give the farmer more control over his field.

While new genetic materials and other technologies can transfer some of the conditions of laboratory control to the farmer's fields, the control that a biological researcher has over social variables is not transferable. It is a control that would reside in an absolute power over resources which the farmer simply does not possess. The scientist does not have to make a profit, and if there is an argument with the kids, it does not threaten the future of the lab. The scientist "controls" certain environmental variables, such as long term accumulation of pollutants, by restricting the time frame of a research project; but this kind of control is not meaningful to farmers, or to those who bear the costs of environmental externalities. Reductionism, thus, influences research in that the scientific capacity inherent in a given laboratory may restrict inquiry into the variables that would be most crucial to research in pursuit of alternative values. To the extent that agricultural research institutions are heavily invested in laboratories that restrict the influence of environmental or social variables, they are limited in their capacity to do research that would serve social and environmental values.

4. FARMING SYSTEMS AS AN ALTERNATIVE FRAMEWORK FOR AGRICULTURAL RESEARCH

Critics who feel that agricultural science neglects alternative values because of its reductionist research techniques have sometimes called for "holistic" research. Holism is presumed to be the opposite of reductionism, so if reductionism is the problem, holism must be the answer. Holism fails as a theory of research choice, however, because it seems to say, "Do everything, and do it now." Even if we could make clear choices in light of the holistic view, holistic management of reductionist research programs is hardly a solution to the problem of revising research agendas in light of existing capacity. If holism is to be an antidote for reductionism, then holism must be a component of the actual research methodology that a scientist employs.

Farming systems research is one name that has been given to research techniques that attempt to resist the evils of reductionism in conducting agricultural research. Some farming systems work makes conceptual links with general systems theory, which strives for holism by "always treat[ing] systems as integrated wholes of their subsidiary components and never as the mechanistic aggregate of parts in isolable causal relations" (Laszlo, 1972, pp. 14-15), but the main methodological innovation of farming systems research is to form a team of researchers that interact extensively with farmers. This team is to initiate their research activity with no preconceptions about what the farmer's problems are. By bringing several disciplinary perspectives together in the team's interaction with the farmer, farming systems is expected to be more sensitive to the range of forces that create problems for the farmer. The forces that impinge upon the farmer's activity may not even be primarily agronomic in origin. They may involve family problems, or conflicts with neighbors. They may involve access to credit or regulatory restrictions. The inclusion of social scientists in the farming systems team is supposed to help the team listen more attentively to the farmer's problems, and to be better able to direct their research capacities toward interventions that address the farmer's needs (Flora and Tomecek, 1986).

Any farming systems team will itself have the same sort of intrinsic capacity that has been associated with the idea of a laboratory. The farming systems team is itself a kind of mobile laboratory, consisting primarily in the multidisciplinary expertise of team members. A team with a great deal of expertise may be able to apply agronomic research skills to problems that address family, community or credit problems, but a team that lacks subtlety and imagination may be no better prepared to deal with these problems than the conventional extension service/experiment station partnership that characterizes traditional agricultural research. Indeed, it seems that the holism of farming systems research comes down to three things: (1) a recognition of the way that existing laboratory capacities limit the applicability of research to alternative values; (2) an open-mindedness that includes a willingness to listen to practitioners and to researchers from other disciplines and (3) recruitment of uniquely talented individuals. While these points may be extremely important for the planning and conduct of research, they do not appear to be a particularly deep response to the problem of reductionism, nor do they require philosophical allegiance to general systems theory. The third point in particular raises the possibility that successful farming systems research depends less upon method than upon serendipity.

If this characterization of farming systems is correct, its advantage over traditional research methodology arises from its practitioners' acceptance of two tenets from practical philosophy, and not from deep metaphysical or epistemological views about reductionism and scientific truth. A group of researchers that lacks appreciation of the limiting effects of scientific capacity, or that lacks sufficient openmindedness does not itself have the capacity to do farming systems research. But being openminded and appreciating one's own limitations do not, in themselves, entail that the farming systems team will have the additional capacity needed to formulate agricultural research programs that are applicable to alternative values. Farming systems researchers may accept a commitment to open-ended problem solving, but openmindedness is far short of a capacity to solve problems related to environmental quality, to small-farm survivability, or to distributive justice. Like productivity itself, these goals must be translated into a technological form if they are to become the focus of an applied science. Even when these alternative values emerge as clear features of the problematic situation, farming systems teams can direct research toward these goals only to the extent that the researchers have abilities that are not, in themselves, requisite to farming systems techniques (and, indeed, may not even be capable of translation into technological form).

Farming systems may be an antidote to reductionism in the sense that multidisciplinary teams can approach problems with a more comprehensive model of the physical and social reality in which farmers operate. There is a higher probability that the proximate cause of a farmer's problems will be included as a dependent variable in the farming systems model. There is nothing in farming systems, however, that requires or even allows the scientists to bring research techniques to bear upon broader or alternative values. Indeed, the most coherent way to arrive at a holistic model of the farm is to see it as a production system. There *are* alternatives. The farm might be a system for preserving family identity across generations. It might be a system for producing good work habits and a sense of responsibility to others. It might be a feedback loop in a larger system that ensures that human populations do not exceed biological carrying capacity. It would be possible to do research on how agronomic and husbandry practices do and do not further these goals. Indeed, some farming systems teams will successfully address such goals, but they will do so not in spite of, but because of, their training in applied science.

5. CONCLUSIONS

Busch and Lacy have done survey research in which the various forms of peer review emerge as leading constraints upon agricultural scientists' ability to alter the research agenda in favor of alternative values. Patterns of journal publication, peer review of research proposals, and the tenure and promotion process influence research choice, and serve as the main vehicles for objective evaluation of research quality (Busch and Lacy, 1983). The picture that emerges is a buddy system in which membership in the club requires fealty to existing prejudices. Dundon has offered case studies of agricultural scientists who challenged establishment emphasis upon productivity that substantially confirm this picture (Dundon, 1986, pp. 39-51). While I have concentrated on agricultural research, many of these points may be generalizable to medicine and engineering.

In the political economy approach, scientists' research goals define a set of competing interests that merge into the polity of science, an interest which is itself but one interest among others. In political economy, it becomes difficult to ask why individual actors are committed to interests, simply because actors are defined in terms of those interests. The possibility of change or synthesis in the defining goals of agricultural research requires individual scientists to change their allegiance from one interest group to another. The resources for bringing about such a change are essentially two: rational persuasion and brute power.

The analysis that has been given here suggests a modification of the political economy described by Busch, Lacy and Dundon. Scientists in applied disciplines are not, in an important sense, free to change their allegiances in response to either rational persuasion or brute power. The technological character of applied disciplines entails that what these disciplines are primarily about is the discovery of means to achieve certain ends. Scientists are equipped for this task in a dual sense. First, their laboratories are literally equipped (and the research staff is trained) to produce discoveries of a certain sort. Second, the discoveries they can and do produce are made legitimate and important because of their contribution to the achievement of the ends to which they are means. Technological values such as productivity become part of the equipment of a research laboratory in the applied sciences in the second sense. Changing to new goals is constrained in the same fashion as changing an assembly line from guns to plowshares. First, the actual machines and skills must be changed, for the technology for one task is not the technology for another. The second level change in technological values is more difficult, however.

Changing an automobile assembly plant into a center for electronic communications might be a more apt analogy. It may appear at first glance that what's needed is a change in material technology, rather than technological values. The facility changes from one organized around the goal of moving people to things, to one of moving things to people. One imagines that little more than the building shell will remain should such a change actually take place. Not only the machines, but the people will be replaced. Changing material technology is no more adequate to reorient research than to reorient a manufacturing organization. In point of fact, the past decade was a time in which the material technology in agricultural research has undergone a profound transformation. The introduction of gene transfer technologies has completely transformed the material technology of agricultural laboratories in virtually every discipline. What has survived is the idea that agricultural disciplines are organized around the conceptual paradigm of solving production problems. It is only insofar as environmental or health impacts have been capable of being portrayed as production costs that agricultural scientists have been able to respond to the interests aligned in support of these goals. Agricultural scientists just do not know "how to do it" with respect to ecology, sustainability, distributive justice or broad scale rural development. The technological values that make the applied science laboratory what it is cannot be changed without substantial reconceptualization of the disciplinary foundations of the given science. What is more, it is not clear that values such as environmental quality and small-farm survivability, let alone distributive justice, are amenable to the practical rationality that can be employed in pursuit of productive efficiency or human health. It is, in part, because contemporary concepts of practical reason present robust, means-end models attached to the goals of food availability and health that we have developed applied sciences around them. The technological character, in other words, of the foundational values is what makes them work as applied goals for researchers in medicine, agriculture and engineering.

An applied scientist is *not* simply a scientist given a specific practical goal to which science can be applied. If this were so in any genuine sense, organizations dedicated to applied science and engineering would serve a much broader array of social goals than they currently do. An applied scientist gets no credit for having done scientific work when the social problems that get solved are of the wrong sort. Elaborate organizational structures have been constructed to decide and enforce the judgment of which problems *are* the right sort. Perhaps this is a good thing. I have said nothing which would seriously challenge the notion that the applied sciences are appropriately configured as they now stand; but the configuration of these sciences is a philosophical issue that deserves consideration in light of ethical standards and the requirements of justice. Without such consideration, we are technologically committed to means that determine our achievable goals, regardless of the goals' relative importance or defensibility.

Texas A&M University

NOTE

[1] This paper has been improved immensely after comments by Phil Shepard, Jeff Burkhardt, and Carl Mitcham.

REFERENCES

Altieri, Miiguel (1987). *Agroecology: The Scientific Basis of Alternative Agriculture.* Boulder, CO: Westview Press.

Busch, Lawrence and William Lacy (1983). *Science, Agriculture and the Politics of Research.* Boulder, CO: Westview Press.

Callicott, J. Baird (1990). "The Metaphysical Transition in Farming From the Newtonian-Mechanical to the Etonian Ecological." *Journal of Agricultural Ethics* 3 (1), pp. 36-49.

Dahlberg, K., ed. (1986). *New Directions for Agricultural Research: Neglected Dimensions and Emerging Alternatives.* Totowa, NJ: Rowman and Allenheld.

DeWalt, Billie (1988). "Halfway There: Social Science in Agricultural Development and the Social Science of Agricultural Development." *Human Organization* 47, pp. 343-53.

Dundon, S. (1986). "The Moral Factor in Innovative Research." In *The Agricultural Scientific Enterprise,* ed. L. Busch and W. Lacy. Boulder, CO: Westview Press.

Flora, C. B., and M. Tomecek, eds. (1986). *Farming Systems Research and Extension: Management and Methodology.* Manhattan, KS: Kansas State University Press.

Laszlo, Ervin (1972). *The Systems View of the World.* New York: George Braziller.

Latour, Bruno (1988). *Science in Action.* Cambridge, MA: Harvard University Press.

Thompson, Paul B., and Bill A. Stout, eds. (1991). *Beyond the Large Farm: Ethics and Research Goals for Agriculture.* Boulder, CO: Westview Press.

M. HEYBOER

THE NORMATIVE IMPLICATIONS OF THE CONFIGURATION OF THE APPLIED SCIENCES: A Comment on Thompson

Thompson's paper examines the feasibility of intervening in the applied science research process to direct it toward the production of alternative technologies. Specifically, Thompson focuses on intervening at the stage of research choice. He proposes a modification of the political economy of research choice proposed by Busch, Lacy, and Dundon, who argue that research choice is influenced primarily by social, political, and economic factors. Thompson argues that technological values (such as productivity), practical rationality, and laboratory facilities also influence research choice, and that these severely constrain the ability of researchers to pursue alternative goals that flow from alternative values. Those alternative values may not be translatable into technological values amenable to practical rationality. He also thinks that the current technological values are at the core of applied science and enable applied scientists to satisfy society's goals. Changing those values would therefore require a substantial reconceptualization of the foundations of the various applied science disciplines. Thompson's argument is important because it forces us to consider the normative implications of the configuration of the applied sciences. His emphasis on technological values, practical rationality, and laboratory facilities also nicely complements the political economy approach to research choice.

Thompson seems to think that the role of social factors is severely limited. He presents the technological values as an independent alternative mechanism shaping knowledge development in applied science through their influence on research choice. And if technological values are as essential to applied science as he claims, then they would limit the influence of social factors on knowledge development. It is unfortunately not clear whether Thompson thinks his analysis of the role of technological values represents a way to

Joseph C. Pitt (ed.), New Directions in the Philosophy of Technology, 153-157.

circumvent the social construction of knowledge, although his discussion does appear to imply that conclusion.

But Thompson's argument ignores the fact that technological values, like all values, are themselves socially embedded and socially constructed and reconstructed. Productivity means different things in different contexts, and thus motivates different kinds of research in those different contexts. And those different meanings, those different kinds of research, and the different uses to which that research lends itself all reproduce the social contexts in which specific technological values are embedded.

Furthermore, Thompson implicitly constructs a hierarchy of values, with technological values at the top of the hierarchy as the dominant values shaping applied scientists' research choice, while other values are subordinate and become important only if they can be translated into technological values or if they can be related to technological values. Undoubtedly, technological values are deeply entrenched in applied science. But it is questionable whether that means that those values are always the dominant values shaping research choice regardless of the context in which applied scientists work – it is even questionable in an institutional context that rewards applied scientists for particular kinds of research and results because that institutional context is part of a much larger context. And it is also not clear that all scientists always operate with a hierarchy of values. It is possible that some scientists in certain contexts operate with a hierarchy of values. It is possible that some scientists in certain contexts operate with non-hierarchical, holistic value systems that accord equal weight to multiple values.[1] In their case, research choice may well be motivated by multiple values.

Thus, as a result of ignoring the social context, at the very least Thompson's argument about the role of technological values is seriously incomplete. To understand the normative issues associated with research choice, his argument needs to be integrated with the political economy approach, rather than merely complementing it.

Thompson's argument about the role of technological values is also in part a political argument. He argues that the applied sciences are characterized by an internal trajectory, due in part to technological values and their necessity, but also to the potential and possible impossibility of translating alternative values into technological values amenable to practical rationality in contemporary research sites (e.g., laboratories). He thereby transforms technological values into a source of power because they limit the possibility for democratic control and guidance of applied science. In addition, he argues that technological values limit the freedom of applied scientists. This implies that technological values limit their accountability because, after all, what choice did the applied scientists have except to accept the imperatives of the technological values? Limiting applied scientists' accountability is also a source of power because it absolves applied scientists from responsibility for the normative implications of their practice. Thompson's analysis thus draws a sharp distinction between the production of knowledge (or technologies) in applied science and the use of that knowledge, and between ethics and knowledge (or the epistemology of applied science). Ethical issues consequently are related solely to goals rather than to goals and means.

The transformation of technological values into a source of power is also due to the failure to recognize the historically and socially constructed and contingent nature of those values. Everyone's freedom is limited, but that does not mean that we have to accept any and all limitations on freedom. We can analyze the sources of those limitations and seek to introduce change. In this case, that would mean examining the social (and historical) contexts in which certain meanings of the technological values emerged, how and why those meanings became entrenched, and especially how and why they have been and are recreated over time. Rather than being forced to accept the normative implications of the technological values that shape research choice, understanding the social relations that sustain and recreate particular meanings of the technological values opens up the possibility for democratic control and guidance of applied science by changing the social relations in which those and other values are embedded in order to achieve desired normative consequences. Introducing social changes will not be easy, should the decision to do so be made, and the question of how to introduce specific social changes looms large. But that does not undermine the point that Thompson's transformation of technological values into a source of power and his demarcation between knowledge development and use is unwarranted due to the fact that technological values emerge out of and are embedded in social processes.

The attempt to absolve applied scientists from responsibility is similarly unwarranted. Scientists themselves can question the normative implications of their practice, even though they are not completely free. No one and no action is completely free, yet it is still reasonable to ask people to consider the implications of their actions and it is reasonable to hold people accountable for the consequences of their actions. The discussion of technological values merely serves to mystify the fact that the normative consequences of applied science can be debated and controlled. At the very least, applied scientists can question the normative implications of both the goals society pursues and the means the applied scientists provide society to achieve its goals. Furthermore, justifying the normative implications of applied science research could become part of the research process, including choosing to conduct particular kinds of research – applied science itself could become explicitly normative rather than pretending that a clear, impermeable boundary exists between knowledge and ethics that insulates applied scientists from the normative implications of their work. Applied scientists could develop such justifications themselves, and they could also open up the process of developing those justifications to democratic participation. This too would open up the knowledge development process to democratic control and guidance by negating the ability of applied scientists to hide behind some supposedly special and essential value system. And since specific values and specific meanings are contextually contingent, negating applied scientists' ability to hide opens up the possibility of democratic control of the processes whereby those values are constructed and reconstructed to steer those processes in particular directions.

Finally, Thompson argues that applied scientists "just do not know 'how to do it'" (p. 14) with respect to alternative values. This, too, would undercut the possibility of democratic control and guidance of research choice. But his argument is unwarranted because it rests on a historical fallacy. Just because applied scientists have not yet been able to incorporate alternative values in the knowledge development process does not mean that they will not be able to do so in the future. It merely indicates that we might need to look deeper into the processes whereby the current values are maintained.

Nonetheless, Thompson's focus on technological values and practical rationality underscores the continuing relevance of epistemological accounts.[2] This is particularly important with respect to applied science because, unfortunately, the epistemology of applied science remains virtually unexplored. The problem of the normative issues associated with the use of science, and the role epistemological values play with respect to those normative issues, should both become matters of continuing study and debate. Yet this problem requires cross-disciplinary and interdisciplinary analysis. Hopefully, we can draw not only philosophers, historians, sociologists, and the Science and Technology Studies community, but also practicing applied scientists into this dialogue.

State University of New York – Institute of Technology

NOTES

[1] See, e.g., Tibor R. Szanto, "Value Communities in Science: The Recombinant DNA Case," in Thomas Brante, Steve Fuller, and William Lynch, eds., *Controversial Science: From Content to Contention* (Albany: State University of New York Press, 1993), pp. 241-63.

[2] On that point, see also David K. Henderson, "On the Sociology of Science and the Continuing Importance of Epistemologically Couched Accounts," *Social Studies of Science* 20 (1990): 113-48.

PAUL B. THOMPSON

REPLY TO HEYBOER

These comments raise a number of legitimate points about the relationship between the argument I have offered in "Technological Values in the Applied Science Laboratory," and the emerging political economy/social constructionist school of thought in the philosophy of science. I think of myself as a cautious proponent of that school, rather than an opponent, and would refer readers to *Beyond the Large Farm: Ethics and Research Goals for Agriculture* (Boulder,CO: Westview Press, 1991) and to *The Spirit of the Soil: Agriculture and Environmental Ethics* (London: Routledge, 1994) for more complete discussions of my views on the philosophy of applied science. For the time being, responding to these comments provides a good opportunity to position my original paper within a network.

The key point regards the interpretation of an inside/outside distinction, the misreading of which has led the commentator to critique a view I do not hold. It is fair to say that, until recently, philosophy and sociology of science have presumed the existence of abstract and indefinite but authoritative epistemological boundary rules that demarcate science proper from non-science and extra-scientific activity. Although demarcation was always viewed as problematic, the practitioners of this old-style philosophy of science generally took scientists' links to defense ministries, pharmaceutical companies, regulatory bodies, members of Congress, the Rockefeller Foundation, the World Bank, the AMA, and so on as extra-scientific. Thus emerged an interpretation of inside and outside that sought to isolate or distill the essence of science proper (the inside), by marginalizing, excluding and ultimately denying the relevance of any environment beyond this epistemological skin.

The interpretation of inside and outside that I offer is far more prosaic. The inside/outside borders to which I refer are those that someone untrained in science, philosophy or sociology would recognize. Laboratories are places, usually enclosed by bricks and mortar, though sometimes by glass or barbed wire, and my sense of when one is in or out of them relies on the determination that any

Joseph C. Pitt (ed.), New Directions in the Philosophy of Technology, 159-162.

child would make. This may not be a strictly literal reading of "inside" and "outside," but people can tell when they are in the lab and when they are out of the lab and in the hall, the restroom or the ballroom of the Palmer House Hotel. Clearly a great deal of what both purists and political economists want to classify as science goes on in these other places. Far from wishing to advocate an acontextual philosophy of science, I wish to examine specific contexts not as theoretical abstractions, but as actual places. I submit that political economy must heed the lessons of political geography.

I *will* defend the claim that what goes on inside the places we call applied science laboratories is dominated by technological values, but I do not think that this claim entails that science (even applied science) is always or even usually so dominated in total. I fully accept the suggestion that a complete normative theory of research choice requires the political economy approach, and indeed I have presupposed such an approach even in my earliest work. As such, I reject the claim that my argument "ignores the fact that technological values, like all values, are themselves socially embedded and socially constructed and reconstructed." The technological values in applied science laboratories are technological because they order activity so as to develop products that succeed or fail to the extent that they produce results characterized by instrumental rationality: a cure for a disease, a material meeting certain specifications, or a plant that produces all of its fruit within a narrow time range. They are values, however, because the laboratory embodies a network of social relationships that extends well beyond its walls.

One can (and many of us do) conduct an ethics seminar inside a laboratory, just as one might (and many scientists do) negotiate a contract, write a manuscript or play bridge there. Technological values dominate these other uses of laboratories for three mutually reinforcing reasons. First, the equipment and human capability inside laboratories can be brought to bear upon the production of novel tools. While new tools or solutions to problems might be worked out in many places, laboratories have a geographical advantage over most alternatives that is attributable to the spatial contiguity of equipment and human capability. Of course, this advantage exists only for a limited range of problems that will be specific to each laboratory; that is one of the main points I wanted to establish in the original paper. Second, the people who control access to and from applied science laboratories (in the sense that I control access to my office or my home) are committed to technological values: they generally believe them to be important and worthwhile. At the least their incentives and opportunities are closely tied to the production of tools and the solutions of problems. Third, the enormous capital investment that is required to equip and staff laboratories makes the people that control capital loathe to abandon functional laboratories. To the extent that the administrators and agencies who supply funds think that it is important to get a return on the capital already frozen in a laboratory's technological capability, resources continue to flow into existing laboratories, even when their products are open to criticism.

Each of these sources of dominance, and especially the last two, are social. Technological values have tremendous influence over research choice, but the source of their power does not originate inside the laboratory. Many of the forces that determine what, if anything, happens inside laboratories are spatially assembled outside the laboratories. Once these forces coalesce in building and staffing an actual laboratory, however, the technological values that dominate the laboratory will resist their realignment. This implies, as the comments suggest, a constraint on the extent to which research choice can be made more democratic. It also implies that we should hardly expect a virologist to transform her laboratory into a space where research on attitudes toward sexuality is conducted, even when the virologist in question becomes convinced that the latter is more vital to suppression of the disease that the lab has been organized and funded to control. The virologist should not be held to account for continuing to do work in virology, in other words. This *is* a limit on the accountability of the applied scientist, as the comments suggest.

But technological values limit the broader accountability of applied scientists far less than they limit the ability of applied scientists to change what they do inside the laboratory. I agree with the claim that applied scientists can and should question the normative implications of their practice. Whether this questioning must take place inside the laboratory is a different and more difficult question to resolve. Certainly it will not be adequate to confine ethical deliberation to laboratories, since the walls of the laboratory will then preclude precisely the sort of public, democratic discussions that are necessary for changing the ensemble of forces that cause laboratories to be constructed according to a given regime of technological values, in the first place. The comments, thus, usefully bring out the implications of my analysis for the accountability of scientists, but that was not the main point of the paper.

I wrote the first draft of "Technological Values in the Applied Science Laboratory" in 1989. It was focussed more narrowly on agricultural science, and was presented at the AAAS meetings in 1990. At that time, agricultural science administrators had not completed (though they were well into) a massive investment in molecular biology laboratories. The original paper was intended to reflect upon an earlier generation of administrators' investment in laboratories that were incapable of responding to critics of chemical and mechanical technologies for agricultural production. Those new laboratories have now been built and staffed, and they will either determine what agricultural experiment stations do for the next few decades, or the enormous financial, political and emotional investment in their construction will simply have to be abandoned. One role for normative philosophy of science is to develop the argument for abandonment. Another is to undertake questioning and deliberation within the constraints of this investment in technological values, but the latter task requires that we understand the constraints, partly by asking how (and if) they function in other areas of applied science. That is what I took the final draft of this paper to be about.

I fail to see how any of my views (in this paper or stated elsewhere) draw a sharp distinction between the production and use of knowledge, or between ethics and knowledge, or entail that ethical issues are "related solely to goals rather than to goals and means." I do not see how anything I have ever written can be interpreted as an attempt to absolve or isolate scientists from ethical responsibility, nor can I recall any time at which I said or implied that the limitations entailed by technological values are anything other than historically contingent and potentially controvertible by future events. I have been a lifelong fan of contingency, and a faithful enemy of the *a priori,* and I have the bruises to prove it. I am, therefore, somewhat at a loss in knowing how to respond to these aspects of Heyboer's comments, and must await future enlightenment from my critic.

Texas A&M University

PHILIP T. SHEPARD

TECHNOLOGICAL NEUTRALITY AND THE CHANGING NORMATIVE CONTEXT OF APPLIED SCIENCE RESEARCH[1]

A popular notion or common picture of applied science sees it as putting basic science to work in solving the world's problems. In this view, while the basic sciences enjoy considerable independence in pursuing research goals that come from the scientist's own discipline and research interests, the ultimate research goals of applied science come from outside of science and the applied scientist can only choose which of those problems to work on and what means of solution to apply to them.

Yet this picture largely ignores the way in which applied sciences, with little or no alteration, have utilized the same modes of professional organization as their more basic sister disciplines, and in guiding inquiry and credentialing newcomers have emulated similar paradigmatic achievements in their choice of goals, methods, strategies and standards of research (in the case of agriculture, see Kunkel 1988, Busch and Lacy 1983). In short, the applied sciences have constituted themselves as disciplines on the basis of the same kinds of models of scientific functioning as the basic sciences and in the process have come to enjoy a level of autonomy greater than that of the hired hand or freelance problem-solver. While somewhat more accountable to societal needs and public direction than practitioners of more basic research, the applied scientist individually and the applied sciences collectively nevertheless retain considerable power over the direction of research.

As Paul Thompson points out (in this volume) the "technological" value of increasing "productivity" in the agricultural sciences has limited their accountability, particularly to new agenda demands for healthier, more environmentally benign, more sustainable, and more socially just agricultural practices and technologies. Scholars of the new sociology of science will, of course, be quick to point out that technological values are not self-generating but emerge from specific social and historical contexts through

Joseph C. Pitt (ed.), New Directions in the Philosophy of Technology, 163-184.

processes of construction and negotiation. Thus they are contingent values and could be reconstructed or renegotiated differently, or alternate values could be constructed or negotiated in their place. Detailed case studies of the construction of technological values promise to illuminate such possibilities and reveal prospects for both greater accountability of applied science and greater effectiveness vis a vis alternative research goals.

Yet the issues of accountability remain normative in character – they concern whether and how far scientists *should* be held responsible to societal or public demands and preferences. Whether one wants to raise the level of autonomy of practicing applied scientists or to make them more accountable, the issue takes its meaning as much from its normative and broadly cultural context as from its concrete social and historical settings.

Today this normative context has become unusually convoluted, odd, and even deceptive, and poses to an honest inquirer considerable difficulty in just getting one's bearings. It is for this reason that I think Thompson's paper calls us to reexamine the normative context of claims of autonomy for applied science. Attempts to ground claims for or against autonomy – either through detailed case studies or through closely reasoned logical arguments – face a myriad of meanings and entrenched forms of misunderstanding that threaten from beginning to end to vitiate the effort. So I want here to back up and reconsider the sort of justificatory arguments that have, in recent times, supported claims for scientific autonomy. My aim is not to avoid but rather to prepare for more penetrating examinations in detailed case studies of the processes by which scientists' power over the direction of research is exercised. I want to supply, in other words, a sort of propaedeutic that would allow honest inquiry to escape from the tangle of loaded assumptions that presently characterize debate over scientific autonomy.

1. THE ODD STATE OF DEBATE OVER TECHNOLOGICAL NEUTRALITY

In recent times justifications of autonomy for applied science have most often been drawn from notions of technological neutrality. Since controversy over various technologies heated up in the 1960s, however, critics of particular styles and choices of technology have mounted increasingly forceful challenges to neutrality concepts (e.g., Winner 1977, 1981; Dickson 1974). Within subfields of Science and Technology Studies, the challenges have been so widespread that today few if any voices are raised in defense of neutrality. And even in the larger domain of academic and scientific scholarship explicit defenses seem scarce.

Yet the lack today of overt theories of technological neutrality does not reflect the rhetorical situation encountered, on the ground as it were, when the issue is raised at the level of practice. A great many people seem to go on taking some vague notion of technological neutrality for granted. In my own experience, for example, with undergraduate students in courses on philosophy of technology over the last twenty-five years and with the literature and conduct of agricultural controversy during the same period, those that buy into the neutrality concept feel little or no burden to defend it; while those who would challenge the concept often seem to be intimidated by the social and institutional authorities tacitly arrayed in its support.

Were the concept in question central to common-sense, everyday beliefs, such as the belief in the existence of ordinary material objects, this rhetorical situation would be less perplexing. "Technological neutrality," however, seems not to be derived from everyday practical reason but to have arisen only as part of the intellectual armament of modernism. The survival of knee-jerk acceptance of neutrality calls for explanation, perhaps in terms of deeper aspects of ideological orientation[2] derived from the practices of a modern technological society. Certainly if the felt burden of proof on neutrality issues answered only to the logic of the arguments taken on their merits, then the myriad challenges alone would have shifted the burden to defenders of neutrality. Since this has not in fact happened – at least not in my small corner of the universe – reconsideration of the issues would appear, again, to be warranted.

Against the backdrop of this odd rhetorical situation, the present discussion will review some main arguments for and against technological neutrality in order to reassess prospects for its defense. That disciplines of applied science and their investments in "technological" values remain major forces in our social world is not in doubt. But whether these disciplines should continue to enjoy their present level of autonomy and whether their investments in technological values are coherent and truly justified is very much in doubt, and perhaps this discussion will bring us closer to realizing why.

2. THE LOGICAL EMPIRICIST ARGUMENT FOR AUTONOMY

The concept of the value neutrality of applied science and technology arises most characteristically in the context of arguments for autonomy. Among such arguments those grounded in logical empiricism are often given the most weight. So let us begin by considering a version of the logical empiricist's argument, in particular for the autonomy of agricultural science:[3]

> Like the basic sciences, agricultural science ought to be granted considerable autonomy from social, moral, and political pressures, aims, and values. The technologies they develop ought to be based on science, and since the method of science is neutral with respect to normative concerns, the science-based technologies produced by agricultural research are neutral also. Being neutral, they develop according to an internal logic or practical necessity of their own. So their developers, the applied scientists, ought to be free to follow that logic of development without outside interference. Such outside demands will disrupt the integrity of scientific inquiry, making its results doubtful and the development process more costly and time consuming. Thus the social order can best manage its investment in agricultural research by delegating authority over technology development to agricultural scientists.

The main lines of this argument are fairly clear.[4] It tries to justify autonomy by a logic of development that allegedly stands apart from normative questions because technology (or applied science), like science itself, is neutral with respect to normative values. If things go wrong with technology, it is often said, the fault is not in the technology itself (which is neutral), but rather in what people do with it. This "use/abuse" model of technological responsibility epitomizes the public face of the principle of technological neutrality (see Dickson 1974, p. 16 and ch. 2). Technological neutrality in turn rests on the alleged value neutrality of science, with help from the additional premise that technology ought to be science-based.

3. A CHALLENGE TO VALUE NEUTRALITY

Among the many challenges to this argument for autonomy, perhaps the most important line of attack was outlined more than forty years ago by Richard S. Rudner in the pages of *Philosophy of Science* (1953). Arguing from premises that have usually been accepted by logical empiricists, Rudner countered that the scientific method itself requires normative judgment. Paraphrased slightly, his counter went like this:

> The scientist qua scientist accepts or rejects hypotheses. To do so in a rational and methodological way, he or she must do so on the basis of evidence. But the evidence available at any given time is never sufficient to deductively imply the truth of a general hypothesis. So to accept a general hypothesis, the scientist must make a judgment that the evidence available is strong enough to warrant acceptance. If that judgment is rational, however, it must take into account the possibility of errors and how important the errors may be in a typically moral, political, or social sense of importance. If we act on the hypothesis and it is false, how serious will the consequences be? or if we fail to act on it and it is true how serious will be the consequences of our omission?

By way of example, Rudner compared the hypothesis that a batch of Salk vaccine will immunize people without infecting them with polio to the hypothesis that a batch of belt buckles are free from dysfunctional defects. In the case of the Salk vaccine, he argued, the seriousness of error is very great in moral terms and, accordingly, the scientist should demand a very high level of evidential support before accepting the hypothesis. In the case of the belt buckles, however, very little is at stake, and so a much lower level of evidential support will be sufficient. But in each case the judgment of seriousness is a normative and not an epistemological one,[5] and moreover it is a judgment that is required if acceptance is to be warranted on methodologically explicit grounds. Hence, the method of science is not value neutral; the scientist qua scientist must make value judgments, whether he acknowledges it or not, if his acceptance of general hypotheses is to be scientifically valid, i.e., methodologically well-grounded.

While this is not the place to explore the extensive discussion Rudner's argument spawned (e.g., Levi 1960, Jeffrey 1956), it is appropriate to point out the grave challenge the argument poses to the autonomy of applied science in general and agricultural science in particular. While some (e.g., Levi) have held that basic science escapes Rudner's argument because of the distance or detachment of its hypotheses from the contexts of action, this rejoinder is by no means available to the applied scientist. The hypotheses tested by applied research are typically of just the sort that Rudner used in his examples. Their bearing on prospective actions is fairly clear and direct, and the consequences of error are identifiable in principle.

Indeed, much criticism of agricultural science has focused precisely on failure to anticipate the consequences of its innovations (Busch and Lacy 1983, Johnson 1990). Because Rudner's point directly targets hypotheses about technics, it would seem to leave the autonomy of agricultural science without a leg to stand on. If technology disciplines acquire their purported neutrality by way of scientific method, and that method can be applied to technics only by making normative judgments, then clearly the purported neutrality is specious.

4. BACKING UP THE RHETORIC OF NEUTRALITY

Where does this leave the proponent of autonomy? I want to suggest that although the rhetoric of technological neutrality remains popular, those who would use it are vulnerable if they are not able, on demand, to back up that rhetoric in a cogent way. Some of the current political difficulties of agricultural scientists surely are accounted for by this gap in justification. But if one takes this point to heart and attempts to fill the gap in justification with a revised account of technological neutrality, one then faces a kind of slippery slope on which it appears very doubtful that the neutrality concept can gain sufficient purchase to sustain a defense.

To what can the neutrality theorist ascribe value neutrality? and in what sense? The popular view appears to imply that neutral things can not be coherently evaluated as good or bad in a normative sense, i.e., not good or bad morally, politically, socially, or environmentally. The use/abuse rhetoric seems to require as much. When the critic lodges a complaint against some form of technic, the answer is, "It's not the technology itself that's at fault, but rather what people do with it." The clear implication is that there is some level of technology that is deeper, more fundamental, or perhaps more abstract than the use of a technic and at that level the "technology itself" is neutral, i.e., neither good nor bad, and therefore not at fault.[6]

This last inference underscores the thrust of the use/abuse model to exonerate technics (artifacts) from blame or from being implicated in moral, social, political, environmental, or other problems that are essentially normative. Technics, on this scheme, have to do only with practical necessity or with problems that are purely technical, purely a matter of means. But if people's use of technics is not exempt from fault finding, how can "technology itself" be exempt? and what is this "technology itself"? My contention is this: if we start with the most concrete level of the being or existence of a particular artifact and work our way (down the slope as it were) to progressively deeper or more abstract levels, we shall find all along the way that normative concepts can be and, indeed, are coherently applied to "technology itself."

5. FAULTY ARTIFACTS

The first step is to look at the technic used as distinct from its use. The neutrality theorist wants to claim that technics qua artifacts or qua physical objects are not at fault in the problems of their use, i.e. not to blame. For example, "Handguns don't kill people, people do." Here, presumably, "blame" is understood as appropriately ascribed only to moral agents, entities capable of intentional action. Since artifacts don't intend to do things, they are blame free.

This argument is clear and valid as far as it goes. The neutrality theorist is quite right to enjoin us from ascribing intentions in any literal sense to artifacts. The problem lies rather in the tacit premise that the only way artifacts may be at fault in normative problems is by their having intentions or being themselves the agents of actions. What this hidden premise misses is that artifacts have a definite character, one that is largely the result of the ways they *embody* the design of their maker. An artifact, by definition, is a made object; and it is created (more or less) according to the plan or design of the maker. Thus it comes to have a definite character – a shape, a material make up, etc. – that more or less answers to the maker's plan. Its character, however, delimits to some extent the scope of uses to which it may be put, i.e., it will not be usable for everything and anything. An ordinary claw hammer, for example, is usable to pound a nail into wood, or perhaps even into concrete, but not into thick steel. Associated with each hammer is a range of intended uses for which it was designed, and the range of intended uses differs for different types of hammer. A ball-peen hammer and a sledge hammer thus have few if any intended applications in common.

Of course a hammer can be used in ways other than those for which it was designed, say as a weapon or as an example in a philosophical argument. These, we might say, are improvised uses. Other uses may be accidental. But even the widest set of possible uses of the hammer (intended or otherwise) are delimited by its character. Improvising, you can break a window with a claw hammer to gain entry to a locked house. Accidentally, you can break your thumb too. But you can't, by accident or improvising, chop down a large tree with a claw hammer or move a planet out of orbit. So the horizon of possible uses of the artifact is delimited by its character, including its designed character and also the other traits that it may incidentally come to have through its production or past use (e.g., a rusty hammer, a broken hammer, etc.)

Now the sense in which an artifact may be implicated in a normative problem is that its character may cause or help to cause the problem. The type of handgun known as the Saturday Night Special, for example, is reputed to be good for shooting only fairly large objects at close range, say another person, perhaps an intruder. So its ready availability has been alleged to contribute to the problem of people killing other people. Unlike, say a hunting rifle, the Saturday Night Special has no other significant use. Its availability might thus be said to anticipate and encourage that people will use it to kill other people. As Richard Weaver suggested (1970), its existence as a deployed weapon makes it a "force in being," and thus not neutral with regard to killing people. Of course, the gun's deployment does not actually *make* anybody use it, to kill or otherwise; but it could still fail to be neutral with regard to killing, in the weaker sense that it is a contributing causal factor which, if removed, would make the killing less likely (though certainly not preclude it).

The way the Saturday Night Special is implicated in the problem of people killing people is fairly straightforward, since it arises from the designed character of the gun. But there are other possibilities as well. For one, omissions in the design of guns, say of the safety catch, could be implicated in problems of accidental shootings; similarly, the omission of engine exhaust filtering systems, such as catalytic converters in automobiles, could be implicated in problems of urban air pollution.

So there are a variety of ways that artifacts may be at fault with respect to normative problems without supposing the artifacts to be moral agents themselves. And most of these faults, as the neutrality theorist will be quick to point out, are derived to some extent from the character of the artifact's maker as moral agent. It does not follow from this, however, that the fault is "really" or "only" in the maker and not in the artifact. Guns without safety catches *are* less safe and from the standpoint of avoiding accidental shootings that's a fault in the gun, as well as in the gun's maker. A remedy which only tries to alter the gun maker's design to include safety catches is incomplete; from the standpoint of avoiding accidental shootings we must get rid of the faulty guns also.

6. THE SLIPPERY SLOPE OF VALUE NEUTRALITY

Let us then pursue our examination of the neutrality theorist's prospects. What "technology itself" could be value neutral? To bring the discussion into the context of agriculture and to see how the slippery slope arises (Figure 1 below), consider the case of the farmer who buys a new John Deere tractor, takes it back to the farm, fills the gas tank, and starts it up. The engine sputters, catches, and then immediately blows up. If our farmer has done nothing to abuse the new tractor, she could reasonably conclude that the motor was bad. She might even say, "It's a lemon." Clearly its performance so far is sufficient grounds to warrant taking it back to the dealership and demanding repair or replacement. So, clearly, at least one case of this model of John Deere tractor had something bad about it.

The neutrality theorist might object here that while one instance could be bad or good, the specific engine model, qua technology, is neither good nor bad. We might even suppose that the vendor takes the same line of defense and insists, first, on repairing the engine. But now what if, when our farmer takes it home again, it does exactly the same thing. Bad luck perhaps? So suppose it gets fixed a second time and then blows up yet again. At this point our farmer is likely to make a simple induction by enumeration and suggest to the dealership that further repair is pointless. The dealership, in response, might concede that something went wrong with the quality control at the manufacturer's plant, yet still insist that there is nothing wrong with the engine model itself.

Let us suppose the dealership replaces the engine with another of the same model. Our farmer takes it home and it blows up again in the same way. By dint of considerable expense of time, energy, and frustration, she appears to be building a convincing case that there is indeed something wrong with the model itself, with the engine design. Though she may not be able to identify the source of the fault, it is certainly coherent for her to suggest the possibility that the engine design *is* flawed and that consequently the engine as designed is useless, dangerous, or both. In short, the model of tractor engine could indeed be a bad one in the sense that it doesn't work, or is dangerous, or the like.

To this argument, the neutrality theorist might respond that so far the badness involved is only technical badness. The problem is either that the instance of the technic is not up to specifications or that the specifications themselves are technically faulty, i.e., engines built according to those specifications don't work right.

The rejoinder has some merit. The engine is indeed technically bad in a sense that has something to do with the engine specifications. But why suppose that the badness is "only," "merely," or "purely" technical? Certainly the consequences for our farmer could be quite serious – in economic terms, if the window for planting a harvestable crop were lost due to the hassles over the faulty tractor; or in moral terms, if one of the engine blow-ups maimed the farmer herself. Only in some cases are technical failings insulated from larger consequences, and to determine whether such a case is at hand requires a normative judgment as to the seriousness of the risks or consequences, much as in Rudner's argument.

If normative judgments can be applied coherently to specific models of a technic, or to the seriousness of the consequences of their technical shortcomings, it is not hard to see that similar evaluations can play an important role in deciding, for example, whether to farm using tractors (of some kind or other) rather than, say, with a horse and plow – in deciding, that is, what use-types of technic to employ in farming. Clearly it would be inappropriate to farm with a tractor under a variety of conditions: if there are no roads to get it to the site of use, if the land is too hilly or rugged, if fuel is too expensive, if maintenance know-how or parts are beyond reach, if people at the farm village are scared to death of combustion engines, if the tractor itself costs too much, and so on. But such judgments are normative, they are not purely technical judgments and may not be technical judgments at all.

What goes for farming with a tractor also applies to integrated systems of technics, e.g., tractor plus implements, or even to the larger sociotechnical system, such as the farming operation that employs the tractor-plus-implements and the farmer as well. Indeed, if we think about it in this way, the business of designing equipment for farming would seem to be replete with norms for judging the adequacy of designs, norms that refer somehow to the potential user or context of use, including the natural and social environment. Thus there appears to be no real hope of anchoring the neutrality claim anywhere on the slippery slope at the level of technics, anywhere, that is, from a particular instance of a technic all the way down to sociotechnical systems. (See Figure 1.)

All of which is not especially surprising, for the neutrality theorist may well have seemed from the beginning to be referring to disciplines of technology or their technological values rather than to technics. But if we turn now to the next level of the slippery slope, we will not find the neutrality theorist's prospects noticeably improved. For, first of all, if designs can be judged normatively then so can design principles and methods. The contexts of agricultural science illustrate this fact amply. Not only have the direction and goals of agricultural research been criticized in recent years, but also, and often explicitly, the methods, principles, and standards that shape the research and development process itself.

A striking example of criticism of the methods of development research arose from the Southern corn leaf blight in 1970 (Doyle 1985). During the fairly wet growing season that year, a new mutation of corn leaf blight spread extensively through the Southeast and into other parts of the country. Virtually all strains of corn under cultivation (hybrids) were affected, as the National Research Council later reported (Horsfall 1972), because all were developed from the same intermediate breeding stock containing T-cytoplasm, which the fungus attacked. While breeders were quick to develop immune alternative stocks and thus avoid a potential catastrophe the following year, criticism of the "genetic vulnerability" induced by hybrid corn breeding methods and the concomitant loss of "genetic diversity" have persisted up to the present. (See U.S. Congress,

House and Senate Hearings, 1980; or Shepard and Hamlin 1993, ch. 12.)

Figure 1. *The Slippery Slope of Value Neutrality*

Technics	A particular instance of a technic The specific type of technic The use type of technic Integrated systems of various use types Sociotechnical systems
Technology	Principles and methods of design or development Assessment principles or methods, technological values like "productivity," "efficiency"
Science	Scientific knowledge, laboratory know how Scientific theories, measurement theories Scientific method
Logic	Inferences, semantic assumptions
Philosophy of Language	Admissible language frames
Metaphysics	Ontological assumptions

Design principles for new technics have also been criticized in recent years, and in some cases the criticisms have been accepted by agricultural scientists and their practices revised. Perhaps most notably, design for energy efficiency, for use of alternative renewable fuels, and for renewable energy supply on a decentralized basis have all been introduced within the discipline of agricultural engineering (Stout 1979, 1984). At Michigan State University's Kellogg Biological Station, the principle that dairy systems should be environmentally benign has been utilized in a recent large scale project. Finally, the now well-known California legal suit on the distributive justice of research on the tomato harvester has questioned the desirability of increasing labor efficiency, a traditional goal of mechanization research.

Such questions about design principles for agricultural technics are echoed by challenges and consequent changes in standards of technology assessment ("technological values") in agriculture. How should agricultural scientists judge the technics they develop or refine? The paradigms of agricultural disciplines have developed historically around a central notion of productivity (Dundon 1982, 1986, Johnson and Wittwer 1984, Thompson this volume). But a steady barrage of criticism has led to a loosening of the productivity paradigm. One response has been the emergence of farming systems research, which brings together teams of researchers from different disciplines, sometimes including social scientists, to better hear client-farmers' needs and better anticipate their problems, in part by looking at the farming system as a situated whole. More recently groups of agricultural scientists have coalesced around themes of sustainable agriculture and sustainable development (Beus and Dunlap 1990). Such movements within agricultural science demonstrate responsiveness to normative criticism of productivity paradigms; the responses, however, would be incoherent or unneeded if the technological values of agricultural disciplines were truly separate from the normative sphere.

There is of course yet another level at which the neutrality theorist may attempt to make a stand, the level of science proper – including scientific knowledge, theories, measurement, and methodology. While this is not the place for an extensive re-examination of the issues about science that Rudner raised, a few basic points are in order. First, even if one accepts the claim that scientifically established facts are true independently of the context in which they are used, it remains to discover for any given context of use what the significance of those facts may be. But to interpret and justify the use of facts requires normative judgment. In carrying factual claims from the supposedly detached realm of theoretical knowledge to the arenas of technology development, and hence ultimately of action, the claims become inextricably involved with problems of normative assessment (Shepard and Hamlin 1993, ch. 2).

Take, for example, the controversy over what yields to expect from organic vs. conventional methods of crop cultivation. Some agronomic research has concluded that yields are lower from organic methods than conventional ones. But organic farmers have not been persuaded by these conclusions, and some of the reasons they resist are germane here. Charles Walters Jr. (1975), for example, has accepted the truth of the agronomists' findings but disputed their relevance to comparing conventional with well-established organic farms. His argument is that until fields have had several years to recover from the stress of conventional farming through ministrations aimed to regenerate the soil's ecology and hence fertility, they cannot be expected to produce at higher levels using organic methods. Here, the way one interprets facts depends heavily on a host of auxiliary assumptions and beliefs, some of which are normative. Organic farmers place great importance, in moral terms, on the health of the soil and the sustainability of their practices; so they have reason to hold out for more ecological theories in agronomy and to resist what they perceive as the mistaken implications of superficial experimental results.

Such examples in the context of farming practice illustrate how facts do not speak for themselves, but require actors to give voice to their significance; and what one actor finds significant, another may not. This obstacle to neutrality in action reflects a more general point about the normative positioning of science proper. Even were we to accept that science qua science is value neutral, it would not follow that any application of its results or methods in a concrete context would automatically also be value neutral. Whether or how far neutrality holds up in action depends always and everywhere on the context of action. This point is reflected by the difficulty of blocking Rudner's attack on neutrality in cases of applied science and technology. Whatever the prospects may be for salvaging the value neutrality of basic science, it does not carry over to applied science and technology. Even if basic science is value neutral and technology ought to be based on science, it still doesn't follow that technology either is or should be conceived as itself value neutral.

It would seem then that the prospects for grounding the neutrality of technology at the level of science proper have grown quite dim. Philosophers will likely notice at this point that there are yet deeper levels to which the neutrality theorists might still retreat. The methods and principles of inference on which scientific reason is ostensibly based, the assumptions of logical semantics which undergird them, the choices of admissible language that those assumptions permit, and the yet deeper assumptions of metaphysics that shape analysis of the referential functions of admissible languages, all are involved in the business of science and might be argued to be neutral with respect to normative values.

It is ironic to note, therefore, that even these levels have not escaped criticism in agricultural policy arenas. Indeed, the agricultural research establishment's critics are quite fond of challenging the language, the inferences, and even the metaphysical basis of established research, particularly what they perceive as the persistent assumption of human domination of nature (e.g., Haynes and Lanier 1982, passim). Moreover, agricultural scientists have been stung by such criticism and a few (e.g., Stuart Hill 1979a,b) have responded with earnest attempts to rethink the metaphysical basis of agricultural science. So even at this level normative challenges to the supposed neutrality of scientific reason appear to be capable of both coherence and validity.

The upshot of this exploration of the slippery slope of value neutrality is that defenders of the autonomy of agricultural science who lean on the technological neutrality thesis find themselves today in a true quandary. Nowhere on the slippery slope does it appear possible for the technological neutrality thesis to gain enough support to adequately defend autonomy. In the neutrality theorist's eyes technology is seen as standing above or separate from the normative sphere. The view thus requires a clear line between what is and what isn't value free. Yet at every position on the slope where the line might be drawn, the appearance of a clear demarcation evaporates and the defender is forced down the slope, finally becoming mired in intractable metaphysical controversy. Since the logical-empiricist *raison d'etre* stems in large measure from the hope that science can escape the bog of metaphysics, the neutrality theorist's position at the bottom of the slope would appear to be terminally self-defeating. If the best the neutrality theorist can hope for is endless metaphysical argument, then perhaps it's time to give up the theory.

7. REFLECTIONS

Prospects for defense of the concept of technological neutrality appear, then, to be quite dismal today, particularly in agriculture. Yet the introduction and rapid spread of transgenic biotechnology laboratories, as Paul Thompson points out, "has completely transformed the material technology of agricultural laboratories in virtually every discipline." New questions about the direction of agricultural research are therefore being raised: What are the new labs to be used for? and who decides? In agriculture the consensus since World War II on the desirability of increasing productivity through mechanical and chemical methods of production has been at least partially displaced by shifting the focus on technological means toward recombinant technologies. "What has survived," writes Thompson, "is the idea that agricultural disciplines are organized around the conceptual paradigm of solving production problems." In this way agricultural scientists have clung to technological values as principles for organizing research and their disciplines.

Thompson is surely correct to declare that "the technological values that make the applied science laboratory what it is cannot be changed without substantial reconceptualization of the disciplinary foundations." But what that statement leaves out of account is the extent to which the very coherence of technological values has been unsettled in agriculture by changes in both the normative context of agricultural science and the material means it is seeking to develop. The "robust, means-end models" that have characterized the practical or instrumental rationality of agricultural science during the post-World War II era no longer seem so robust. Many of the public demands now placed on agricultural research are not effectively addressed by those models, in part because new goals go way beyond healthy food and in part because the justifiability of the means by the ends is fraught with ethical complications – as, for example, when resources become more concentrated in a few hands while rural poverty deepens and spreads. The "good means" of the past agricultural era no longer warrant unconditional consent, and on the questions of ends no clear consensus is at hand.

The agricultural sciences, then, are coming face to face with the contingency of their historical foundations. Resting as they do on the slippery and much eroded slope of value neutrality, and having been further dislodged by the introduction of transgenic laboratory biotechnologies, those foundations now teeter on the brink of collapse. Since the agricultural scientists can go neither backward nor forward without reconceptualizing their basis, we should conclude, evidently, that they have no choice but to rethink what they are doing and why they are doing it. And if defensible foundational values do not have a discernible technological character and cannot be translated into particular styles of technology development such as mechanization, then the very character of agricultural science may need to change and its allegiance to an autonomous instrumental rationality abandoned. Instead of seeing themselves as developers of the good means defined by disciplinary conventions, agricultural scientists may need, for example, to revision themselves as more free-wheeling problem-solvers who assemble diverse social and technical resources in new combinations to address problems encountered and who are to be rewarded more on the basis of their success in problem solving than on their contributions to quasi-autonomous disciplines.[7]

The inertia of present institutions of agricultural research, while having brought us to the current impasse, may still seem to moot questions of the autonomy of agricultural science. The scientists may indeed feel blocked from exercising their larger responsibilities, not because they have chosen autonomy, but because they are constrained by their education, discipline, or laboratory facilities. But this is the mootness of the institutionally moribund; it presages not a resurgence of scientists' power over the direction of research but the demise of their institutions and projects. If agricultural scientists want to negotiate an ongoing institutional niche in which they enjoy widespread public support, then non-arguments won't do the job. The changing normative context of foundational arguments for agricultural science must be addressed: good means no longer stand on their own without consideration of relations to ends, the goals of food sufficiency no longer seem paramount over sustainability, justice, or perhaps other goals as well, and the technological horizon has opened to a variety of developmental paths.

What this reconsideration reveals then is the present hollowness of agricultural scientists' investments in technological values. The investments continue to be made, to be sure, but the values that are being bought lack adequate support. The rhetoric that appeals to technology itself as neutral or to good means as good independently of the ends served or the side-effects encountered is hollow rhetoric. Nor do alternative supporting justifications that will withstand public scrutiny appear to be ready to hand. While the impending foundational crisis of agricultural science can be denied or ignored, it does not appear at all likely that it will go away.

Michigan State University

NOTES

[1] Revised from "The Vicissitudes of the Technological Neutrality Thesis in Agriculture," a paper presented at the Sixth Biennial Meetings of the Society for Philosophy and Technology (1991) and published in a preliminary version in Pitt and Lugo 1991.

[2] This concept of ideology is developed in Shepard and Hamlin 1993, esp. chs. 4 and 5. Three orientations prevalent in controversy over agricultural science and practice are described in depth in Part II of that work. Of those, the conventional productivist and the ecological progressive are most likely to endorse technological neutrality in some form.

[3] Essentially the same argument could be used to support the autonomy of all applied sciences – food science, engineering, medicine, and so forth.

[4] Although the main lines of the argument are clear, several points need further explication. Perhaps most salient is the rather loose and ambiguous way the term "technology" is used. Many things are lumped together by this term that perhaps differ significantly. I will sort them into three groups for the sake of this discussion: (1) *disciplines of technology* are organized practices for creating, improving, or assessing ways of provisioning people or social systems; (2) *laboratory technologies* are the equipment, techniques, and routines used in research and development – they may be standardized or improvised, borrowed from available stocks or rigged for the occasion; and (3) *technics* are the tools, instruments, machines, utensils, utilities, devices, biological stocks, routines, and techniques which are the direct means available to provision people or social systems; *provisioning* involves such goals as supplying food, maintaining healthy people, clothing, sheltering, and transporting them, facilitating communication between them, and producing and delivering goods and services for both consumers and industry. There are, of course, definite interconnections and overlaps between the three kinds of thing called "technology."

A second point of clarification concerns what it means for technology to be "science-based." Here again there are multiple meanings. In perhaps the strongest sense, the development of a technic would be science-based if the process made essential use of the scientific method, i.e., if the technic couldn't have been developed without utilizing methodologies of science. The products of development, however, the technics themselves, can be science-based in a weaker sense, if they are produced using available scientific knowledge. Technics that are science-based in this weaker sense presuppose scientific method only indirectly through the use of scientific knowledge. A still weaker, hybrid sense might be countenanced also. Traditional technics, e.g., herbal medicines, might be regarded as science-based if their effectiveness is confirmed by scientific tests.

Thirdly, the logical empiricist argument for autonomy does not attempt to evade ethical constraints on research on human subjects or on products or processes that pose risks to humans. The argument is not proposing that scientists be licensed to produce Frankenstein's monster or to release into the environment truly dangerous transgenic organisms.

Finally, the scope of the argument needs some clarification. Which elements in the planning of scientific research does it include? Typically the argument sanctions scientists' control over the design of their research and their proximate goals or developmental objectives. Does it also sanction control over the more ultimate goals of research? In other words, is the agricultural scientist to be free to set the most basic goals of research or is she in charge only of how to go about accomplishing basic goals that are set by others? On this point there has been much confusion, and some of it is probably due to the difficulty of drawing a clear and stable line between proximate and ultimate goals in research. Yet we need not be hindered by this matter, for neither the logical empiricist argument nor current practices of making agricultural research policy question public determination of the basic goals of research.

The argument for autonomy does not question public determination of basic goals, because it is itself based on a claim of neutrality with respect to such goals. The methodology of science, on which it grounds the neutrality claim, has no bearing on what the ultimate goals of research ought to be. And clearly the "neutral" knowledge supposedly supplied by science does not establish its own purpose or use.

Moreover, the practice of allocating public agricultural research funds is and no doubt will continue to be shaped by public policy. While agricultural scientists participate in the policy process, they do not dictate the outcomes. Certainly in the United States much agricultural research is carried out within institutions with chartered service functions. Many agricultural scientists work within these institutions and they have usually been responsive to the institutions' goals – to client needs, to state and Federal mandates, and so forth. So it is reasonable to suppose here that the thrust of the argument for autonomy extends only to scientists' control over the design and proximate goals of research. The business of setting and interpreting broad goals such as food sufficiency belongs in the public arena and should not be decided solely by agricultural scientists.

[5] As Rudner recognized, valuative judgment is also an essential part of epistemology (the theory of knowledge), but the characteristic evaluations of epistemology concern epistemological values, i.e., standards or criteria for judging the status or grounding of knowledge claims, such as coherence, evidential support, simplicity, plausibility, and so forth. Logical empiricists can readily acknowledge the role of such epistemological values in science; their claim of value neutrality refers rather to non-epistemological values such as those of common morality, politics, and so forth. So the term

"normative" in this paper refers only to non-epistemological values or norms.

6 It is important here not to confuse the broad claim of technological neutrality with claims of neutrality that have very limited scope, or, in other words, with claims that a technic is "indifferent in context" with respect to certain well-delimited, alternative uses. For example, a loaded hand gun (safety catch off) placed on a table at equal distance from two unrestrained, able-bodied antagonists equally skilled in the use of firearms might be said to be indifferent in context to the interest of each in shooting the other. Such examples can be found or concocted in abundance. The problem for the neutrality theorist is that no number of such claims will add up to the broader claim of technological neutrality. Each limited-scope claim of indifference is relative to its context; but the neutrality thesis, at whatever level it is to be construed, must be appropriately general if it is to somehow justify setting "technology itself" apart from the normative sphere.

7 A similar notion of problem-solving research is utilized in Johnson and Wittwer 1984 and elaborated extensively in Johnson 1986.

REFERENCES

Beus, Curtis E. and Riley E. Dunlap (1990). "Conventional versus Alternative Agriculture: The Paradigmatic Roots of the Debate." *Rural Sociology* 55:590-616.

Busch, Lawrence and William B. Lacy (1983). *Science, Agriculture, and the Politics of Research.* Boulder: Westview.

Dickson, David (1974). *The Politics of Alternative Technology.* New York: Universe Books.

Doyle, Jack (1985). *Altered Harvest.* New York: Penguin.

Dundon, Stanislaus J. (1982). "Hidden Obstacles to Creativity in Agricultural Science." Pp. 836-869 in Haynes and Lanier (below).

Dundon, Stanislaus J. (1986). "The Moral Factor in Innovative Research." Pp. 39-51 in *The Agricultural Scientific Enterprise*, edited by L. Busch and W. Lacy. Boulder, CO: Westview.

Haynes, Richard and Ray Lanier, eds. (1982). *Agriculture, Change and Human Values: Proceedings of a Multidisciplinary Conference*, 2 vols. Gainesville: Humanities and Agriculture Program, University of Florida.

Hill, S. B. (1979a). "Ecology, Ethics and Feelings." Pp. 593-607 in *The ReEvaluation of Existing Values and the Search for Absolute Values.* New York: International Cultural Foundation.

Hill, S. B. (1979b). "Eco-agriculture: the Way Ahead?" *Agrologist* 8(4):9-11.

Horsfall, James, G., *et al.* (1972). *Genetic Vulnerability of Major Crops.* Washington, D.C.: National Research Council.

Jeffrey, Richard C. (1956). "Valuation and Acceptance of Scientific Hypotheses." *Philosophy of Science* 23:237-46.

Johnson, Glenn L. (1986). *Research Methodology for Economists: Philosophy and Practice.* New York: Macmillan.

Johnson, Glenn L. (1990). "Ethical Dilemmas Posed by Recent and Prospective Developments with Respect to Agricultural Research." *Agriculture and Human Values* 7(3-4, Summer-Fall):23-35.

Johnson, Glenn L. and Sylvan H. Wittwer (1984). "Agricultural Technology until 2030: Prospects, Priorities, and Policies." East Lansing, MI: Michigan State University Agricultural Experiment Station, Special Report #12.

Kunkel, H. O. (1988). "Issues of Academic Disciplines in Agricultural Research." *Agriculture and Human Values* 5(4):16-25.

Levi, Isaac (1960). "Must the Scientist Make Value Judgments?" *Journal of Philosophy* 57(11).

Pitt, Joseph C. and Elena Lugo, eds. (1991). *The Technology of Discovery and the Discovery of Technology.* Blacksburg, VA: Society for Philosophy & Technology.

Rudner, Richard S. (1953). "The Scientist qua Scientist Makes Value Judgments." *Philosophy of Science* 20:1-6.

Thompson, Paul (1995). "Technological Values in the Applied Science Laboratory." This volume, pp. 137-149.

Shepard, Philip T. and Christopher Hamlin (1993). *Deep Disagreement in U.S. Agriculture: Making Sense of Policy Conflict.* Boulder, CO: Westview.

Stout, Bill Alvin, and C.A. Myers (1979). *Energy for Agriculture,* 7th ed. East Lansing, MI.

Stout, Bill Alvin with Jes Asmussen, *et al.* (1984). *Energy Use and Management in Agriculture.* North Scituate, MA: Breton.

U.S. Congress, House Committee on Agriculture (1980). Hearings before the Subcommittee on Department Investigations, Oversight and Research, Ninety-Sixth Congress, on H.R. 999, a Bill to Amend the Plant Variety Protection Act. Washington, D.C.: U.S.G.P.O.

U.S. Congress, Senate (1980). Hearings on the Plant Variety Protection Act, June 17 and 18. Committee on Agriculture, Nutrition, and General Legislation. Washington D.C.: U.S.G.P.O.

Walters, Charles, Jr. (1975). *The Case for Eco-Agriculture.* Raytown, MO: Acres U.S.A.

Weaver, Richard M. (1970). "Humanism in an Age of Science and Technology." *Intercollegiate Review* (ed. R. Hamlin) 1-2. Reprinted in *Philosophy and Technology: Readings in the Philosophical Problems of Technology*, edited by Carl Mitcham and Robert Mackey. New York: The Free Press, 1972.

Winner, Langdon (1977). *Autonomous Technology*. Cambridge, MA: The MIT Press.

Winner, Langdon (1981). "The Political Philosophy of Alternative Technology." In *Technology and Man's Future* (3rd or 4th ed.), edited by Albert H. Teich. New York: St. Martin's.

RONALD LAYMON

IDEALIZATIONS, EXTERNALITIES, AND THE ECONOMIC ANALYSIS OF LAW

1. INTRODUCTION

What is the logic of technological choice? An elementary first move in answering this question is to distinguish between normative and engineering components. On this view, values from the normative component are used to determine the choices to be made from among the possibilities revealed by science and engineering. But these possibilities–the raw material, as it were, for the normative analysis–are a function of the idealizations and approximations used. Because nothing can begin to happen in the way of testing or application of theory in the absence of some calculated numbers, scientists and engineers require real, as opposed to in principle only, computability. But real computability must make do with actually available empirical data, auxiliary theories, computational resources, and mathematical methods. Given real world limitations on the availability and power of these necessary components, idealizations and approximations must be used by both scientist and engineer. There is really no choice for either practitioner but to simplify. As we shall see, such simplification causes problems for the reliability of the claims of science and engineering. My principal concern in this paper, however, is to show that something akin to idealization, and perhaps approximation, affects the normative component of technological choice as well. After briefly distinguishing and sketching the use of idealizations and approximations in science and engineering, I shall examine idealization in normative theory. In particular I shall focus on the use of idealizations in the way negative externalities are treated in what has become known as the economic analysis of law.

Joseph C. Pitt (ed.), New Directions in the Philosophy of Technology, 185-206.

2. APPROXIMATION AND IDEALIZATION IN SCIENCE AND ENGINEERING

I spoke above of idealization and approximation. What is the distinction? Consider the following two paradigm cases of approximation. The first concerns the ideal pendulum. Its governing equation is:

$$ml\,d^2\phi/dt^2 = -\,mg\sin\phi$$

As usual, m is the mass of the pendulum bob, l the length of the suspension cord, g the gravitational strength, and ϕ the angular displacement. If ϕ is substituted for $\sin\phi$ (since $\phi \cong \sin\phi$), this equation is *transformed* to,

$$ml\,d^2\phi/dt^2 = -\,mg\phi$$

And this equation represents a system whose period is easily calculated to be,

$$P = 2\pi\,(l/g)^{1/2}$$

Our second example deals with the governing equation and initial conditions for a mass projected upward (a distance x) from the surface of the earth (of radius R) with velocity V. The relevant equations and boundary conditions are:

$$d^2x/dt^2 = -\,gR^2/(x+R)^2 \qquad x(0) = 0, \qquad dx/dt = V$$

The first equation can be transformed by substituting 1 for $R^2/(x + R)^2$, (since $R >> x$) which yields:

$$d^2x/dt^2 = -\,g \qquad x(0) = 0, \qquad dx/dt = V$$

And this set has the easily obtained solution,

$$dx/dt = -\,gt + V, \qquad x = -\,(1/2)gt^2 + Vt$$

What's going on in cases such as these? In general terms, we can say that there is the substitution of some term or expression X' for X in an existing mathematical function or equation, where it is hoped that the function is *stable* with respect to the transformation. And here I use *stable* in the ordinary mathematical sense of meaning that if X' and X are "close," then functional performance differences will be "small." Such transformations are commonplace in science and engineering. The motivation for using them is usually the desire to achieve computational tractability, although on occasion they may be introduced for explanatory or pedagogical purposes. Such transformational approximations come in two extreme flavors, namely, those where mathematically derived error bounds are available, and those where such error bounds are not available. In addition, there is something of a continuum between these two extremes. On the mathematically short side, the best one can do is to check for "local consistency." Roughly speaking, terms assumed small in comparison with others should remain so once the transformed equation is solved or the transformed function is allowed to run its course. Such local consistency is generally a necessary but by no means sufficient condition for actual stability. (See Lin and Segel 1974, pp. 188-189.) In between we should expect to find cases that mix physical and mathematical reasoning, where the basic move is to assume, as a working hypothesis, that the stability virtues of the phenomenon studied are possessed by the mathematical treatment as well. Perturbational analysis provides excellent examples of such mixed reasoning.

We now move to a consideration of *idealization*. What comes to mind here are treatments or analyses that utilize point masses, ignore medium resistance, treat finitely sized plates as if they were infinitely extended, or assume the universe has two or even only one massy body. Other idealizations include treating continuous items as if they were discrete, and treating discrete distributions as if they were continuous. As can be seen, the two examples I gave earlier of approximation also contain elements of idealization. What are the stereotypic properties of idealizations? First and foremost, they ignore factors known to be causally relevant. More generally, there is a misstatement of the facts (or what are taken to be the facts) where this misstatement is essential to *launching* or *initiating* a formal analysis or treatment. One simply doesn't know what else to do. Given the absence then of a more realistic treatment, there will be no mathematical bounds on the effect of the idealization assumption. This absence may be mitigated somewhat by experimental evidence that supports the stability of the phenomenon with respect to the idealized or ignored causal factor. In addition, insofar as something is known about the general form of the unavailable superior account, this knowledge may be used (perhaps along with experimental evidence) to support claims about the reliability of the idealized treatment.[1] But in general when things go wrong it will be difficult to determine what should be changed in the analysis.

In summary, we can say that the problem created by the use of idealizations (and unbounded approximations) for science is to determine whether failures to achieve experimental fit to within experimental error are due to the falsity of the underlying theory or of the idealizations used. (Henceforth, I shall use idealization *simpliciter* to refer to both idealizations and unbounded approximations.) In other words, the problem is to determine when we can *praise* theories for achieving as close a fit as is achieved and *blame* the idealizations for the failure to achieve experimental fit to within experimental error. In rare cases where experimental fit to within experimental error is achieved, it must be determined whether this is due to the truth of theory and fortuitously canceling idealizations, or to a fortuitous combination of false theory and false idealizations.

For the engineer the problem seems rather different. Experiment from this point of view reveals the closeness of predictive fit that can be achieved by theory *and* idealization for the particular case examined. If this closeness of fit is good enough for some practical purpose, then the engineer's job is done assuming that he or she restricts his or her practice to the duplication of the experimental case. If predictive fit is not close enough or if there are reasons to consider variation, then the engineer needs to know how well the experimentally determined closeness of fit will *extrapolate* to different cases.

The most common response to the engineer's projectibility problem is the conservative one of minimizing in new physical systems deviations from earlier successful examples. The insight here is simply that because there may be lurking discontinuities, one should not be too venturesome in testing the continuity of known deviations between real performance and idealized analysis. Because these hoped-for continuities typically are not, as we have seen, grounded by theory, "every new engineering design is an experiment, as small departures from convention may have disproportionately disastrous results" (Pippard *et al.* 1953, p. 191). An interesting case where this very general approach of not overly testing continuity was violated is that of the Tacoma Narrows Bridge, "Galloping Gertie." This bridge was considerably more flexible and narrow than existing long span suspension bridges, a fact noted by the board of engineers reviewing its failure.

With a depth ratio of stiffening girders of 1/350, the Tacoma Narrows Bridge with a much smaller weight went far beyond the precedents, notably the Golden Gate Bridge with a depth ratio of 1/168, and the Bronx-Whitestone Bridge with one of 1/209. . . . In respect to width also, the Tacoma Narrows Bridge, with a ratio of 1/72 surpassed in slenderness all others, which show a range of ratios between 1/14 (Triborough) and 1/47 (Golden Gate) (Amman *et al.* 1941, p. 74).

These large variations from existing examples were also explicitly commented on before the disaster in an official review of Moisseiff's design proposal. In addition, general skepticism, but unfortunately only general, was expressed about Moisseiff's argument that existing structural theory justified his design as being adequately stiff.

There seems to be some question even in [Moisseiff's] mind as to whether the obtained stiffness is other than "rather satisfactory." The ratio of width to span is 1/72 which greatly exceeds the corresponding ratios of other long suspension bridges. . . . It therefore seems to me that it would be advisable to widen the super-structure to 52 ft. . . . This width would give a ratio of 1:54 approx. and would provide greater convenience and capacity for highway traffic. The cost would be increased considerably, but the additional cost would certainly be justified (Condron 1938, pp. 4-5).

These general concerns about the large variation from existing practice were rejected, however, by another team of consultants.

It might seem to those who are not experienced in suspension bridge design that the proposed 2800-foot span with a distance between stiffening trusses of 39' and a corresponding width to span ratio of 72, being without precedent, is somewhat excessive. In our opinion this feature of the design should give no concern. The development of the deflection theory of suspension bridge design in recent years for both vertical and lateral deflections has proven beyond doubt that the matter of width ratio is limited not by structural stress but only by the amount of lateral deflection in wind which can be realized without discomfort or fear to the driver of an automobile over the bridge (Andrew *et al.*, 1938, pp. 6-7).

What we have here is an example of another engineering response to the projectibility problem; one argues on the basis of the underlying theory and an enumeration of the relevant physical causes that no discontinuities are to be expected, i.e., that empirically determined closeness of fit and associated corrections can be safely extrapolated in a continuous fashion. Because the Tacoma design called for a simple girder structure, the drag or horizontal forces generated by a heavy wind were expected to be large. Using experimental models to estimate the drag coefficient, Moisseiff and his team designed the bridge to withstand the drag created by winds in excess of one hundred miles per hour. By contrast, the lift forces generated by suspension bridges (given their lack of a streamlined airfoil) "are small in comparison with the dead and live loads and are therefore generally neglected in the strength calculations" (Amman *et al.*, 1941, p. 99). So it seemed safe to ignore these forces, and to accept the "deflection theory" as adequate for the task at hand.

Unfortunately the bridge collapsed. There's a general philosophical research problem here, namely, to determine the ways that engineers attempt to minimize the risk of using idealized analyses or theories.[2] But my task in this paper, as indicated in my introduction, is to illustrate and explore the ways that idealizations affect our normative reasoning.

3. IDEALIZATIONS AND NORMATIVE PRINCIPLES

The thesis that I wish to propose for consideration is that the use of idealizations may introduce risk not only into our analyses of physical systems but also into our analyses of normative problems. Just as a bridge design may look to be safe and secure from the perspective of a particular idealized analysis, so too a course of action may look to be correct from the perspective of a particular idealized normative analysis. A corollary of this thesis is that there is the possibility of the normative equivalent of the Tacoma Narrows bridge disaster. That is, our technological choices might be wrong even though the technology is delivered as promised and even though our normative rationales are unobjectionable in their idealized setting. That is, the incorrectness of our choice would not be because of failed technology, or incoherent or invalid normative reasoning, but because of the idealizations used in our normative reasoning. Such a possibility may seem a conceptual impossibility because of an important difference between engineering and normative argumentation. In engineering (as well as science) there is the possibility of an empirical test. The bridge falls or it doesn't. In other words, we can usually recognize technological success or failure narrowly construed. But there seems to be no normative equivalent of such empirical testing. If this is so, then how could one recognize that a wrong moral choice had been made *because of the use of idealizations*?

One way of dealing with the use of idealizations in science and engineering is to require that predictive output be stable with respect to increasingly more realistic idealizations. Analogously we should seek stability in our moral and social analyses as well. Hence one way that a normative analysis may go wrong is that it be *unstable* given changes in idealization. In such a case, the normative directive will be a function of the idealizations used and not of the facts of the case. And if this is so, then our normative choices will be questionable not only with respect to the underlying values appealed to (the traditional problem of relativism), but also with respect to the idealizations used in conjunction with those values.

It is my impression, as a philosopher of science, that value theorists–those who work in ethics or social and political philosophy–have not paid sufficient attention to the importance and difficulty of demonstrating the stability of normative output given changes in idealization.[3] But this is hardly surprising given how little attention idealizations have been given by philosophers of science. Science itself has been idealized by philosophers in order to come to grips with its complexity. As a general strategy there is surely no harm in this. What does seem wrong, at least from my perspective, is placing to the side as a mere complication what is a central feature of modern science and engineering, *viz.*, the use of idealizations and approximations. The need to bring idealizations and approximations to the forefront is forcibly expressed by Moulines (who describes the problem only in terms of "approximation"):

> Systematic accounts in the philosophy of science seem to have regarded [approximation] as an uninteresting accidental feature of the scientific enterprise-something not really belonging to the "essence" of science and not worth deserving a detailed analysis. We believe this attitude to be fundamentally misleading. It is misleading to take for granted that the canonical situation the philosopher should account for is the situation in which the theory with perfectly exact concepts perfectly fits the facts it is supposed to systematize. There is at present no significant empirical theory in which this really happens, and we see no reason to suppose it ever will happen. An adequate logical reconstruction of empirical science must take approximation as one of its essential features.[4]

My aim in this paper is to show that problems due to the use of idealizations are not restricted to science and engineering. Indeed, I shall attempt to show more, namely, that problems due to the use of idealizations are even more severe in certain realms of normative argumentation.

4. NEGATIVE EXTERNALITIES

I am not going to try to give a general analysis of or argument for the above claims. Instead, I simply want to sketch the rudiments of an example that illustrates and makes tangible the sorts of problem that I am aiming to characterize. The example deals with the way negative externalities are treated in what has become known as the economic analysis of law.[5] Externalities are defined to be the byproducts of one person's (or group's) consumption or production which affect the welfare of others. E.g., noxious fumes from a chemical processing plant. The preferred framework for the analysis of such externalities is that provided by standard welfare economics. The reader may recall that it is a basic assumption of welfare economics that interpersonal utilities are not comparable. In the absence of such comparability one makes do with the surrogate notions of Pareto-superiority and optimality. A quick review of these notions may be in order. If S_i is a possible distribution of goods, then:

> S_1 is *Pareto-superior* to $S_2 =_{df}$ No one prefers S_2 to S_1 and at least one person prefers S_1 to S_2.

S_1 is *Pareto-optimal* $=_{df}$ There exists no S_n such that S_n is Pareto-superior to S_1.

Some simple consequences of these definitions that we shall need (see Figure 1) are that:

1. S_1 is Pareto-superior to S_2 entails that S_1 is greater in utility than S_2.
2. S_1 is Pareto-optimal does not entail that S_1 is a utility maximum.
3. S_1 is Pareto-optimal and S_2 is not Pareto-optimal do not entail that S_1 is Pareto-superior to S_2.

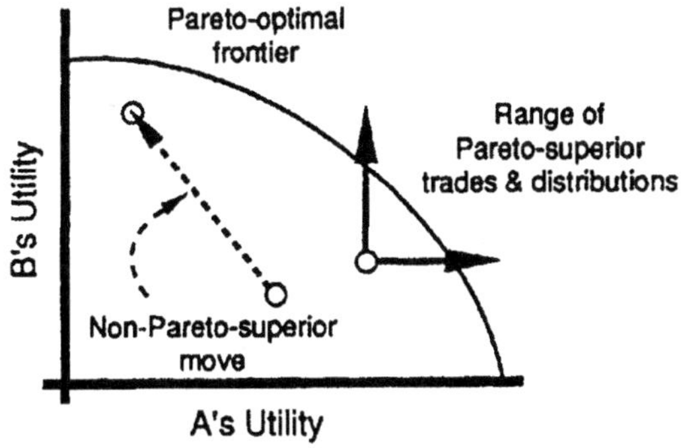

Figure 1: Pareto-Optimality and Superiority

Consider now the problem of adjudicating between a farmer and a rancher whose cattle trample the cornfields of the farmer. Here the damage caused to the farmer's cornfields is the negative externality of cattle production. Because my purposes are primarily illustrative, I shall assume the simple schedule of production and damages given in Table I (Schedule of Production and Damages).[6] The problem of how such a schedule is to be determined can be ignored for present purposes.

In order to get the mathematical formalism cranking, some idealizations are required (just as they are in pure and applied science). So as a starting point, assume a possible or idealized world

	Marginal Profit to Rancher	Marginal Damage to Farmer
steer 1	$50	$10
steer 2	$40	$20
steer 3	$30	$29
steer 4	$20	$40
steer 5	$10	$49
steer 6	$0	$57

Table I. Schedule of Production and Damages

where there are no third party effects, transaction costs are zero, no strategic bargaining is employed. In addition, assume that the problem of adjudication is conceived as one of assigning production rights to either the rancher or the farmer. Consider now the consequences of assigning these rights to the rancher. How many cattle would be produced in such a case? Bargaining between farmer and rancher makes sense for the fourth and fifth steers that the rancher contemplates producing. In the case of the fifth steer, the farmer can rationally offer up to $49 to forgo production. In this case the rancher requires at least $10. In the case of the fourth steer, the farmer can offer up to $40 to forgo production, while the rancher requires at least $20. Bargaining stops at the third steer, for here the farmer can offer at most $29 to forgo production. But the rancher requires at least $30 in order not to take a loss on the transaction. The resulting production schedule of three steers and payments to the rancher will be Pareto-optimal because there is no way to improve anyone's welfare without worsening that of the other party. The situation can be graphically represented as in Figure 2. As can be seen, each point in the range of possibly efficient bargains is Pareto-superior to what the situation would have been if five cattle had been produced. Furthermore, once an agreement has been reached (i.e., once the parties relocate to a point on the range of possible bargains), that agreement will be Pareto-optimal because any future transfers between the parties must worsen the welfare of at least one of the parties.

Consider next the case where production rights are assigned to the farmer. Here, the rancher must pay the farmer for the right to produce his cattle. For each of the first three steers, the rancher will be able to offer the farmer more than will be lost due to damage by the cattle. For example, in the case of the third steer, the rancher will be able to offer up to $30 to offset the $29 worth of damages to the rancher. But in the case of the fourth steer, the rancher can offer at most $20 to cover the expected (marginal) damage of $40. So as was the case when ranchers had production rights, the production schedule when farmers have such rights calls for three steers. Again the resulting allocation of productive resources and payments will be Pareto-optimal. These relationships are graphically depicted in Figure 3.

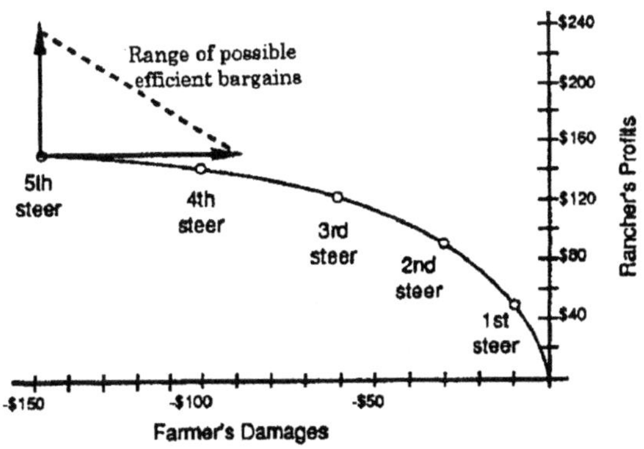

Figure 2: Rancher Has Production Rights

This little example illustrates what has become known as Coase's theorem, which states that in a suitably idealized realm allocative efficiency is independent of the initial assignment of productive entitlements. The production schedule is said to be allocatively efficient because (1) it is a constituent of a Pareto-optimal equilibrium state, and (2) it's the same production schedule that would have been chosen if the rancher and farmer had been a single entity. In this latter (even more idealized) case, the production schedule is once again Pareto-optimal; this because the hypothesized single agent can make no changes in his or her allocation of productive resources that would result in a welfare improvement.

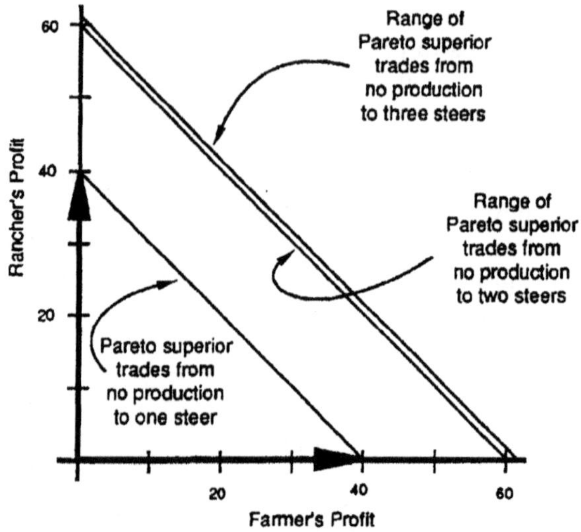

Figure 3: Farmer Has Production Rights

To summarize, in our idealized world, call it w_O, the cattle production schedule is (1) independent of initial productive entitlements, and (2) efficient in the indicated sense. So it follows that overall *utility* will always be increased in the sense that all successful bargaining will result in Pareto-superior transactions. And this regardless of the initial assignment of production rights. Furthermore, because farmer and rancher are assumed to be rational, i.e., to maximize their expected utilities, the bargaining is *by consensus*. And because of this, the autonomy of the agents is respected. Hence, both utilitarian and deontic requirements are satisfied. Therefore, in w_O, there is in either case of rights assignment a harmony between the pursuit of efficiency and the exercise of autonomy. Of course, the assignment of production rights does affect the relative wealth of the agents. (Compare Figures 2 and 3.) Hence, efficiency and consensus by themselves do not determine a rights assignment. Nor do they prohibit *ex post* adjustment or compensation that would move parties to some other location in the welfare space. Assume, for example, that the rancher is given production rights. As we have seen, bargaining will proceed and the rancher will produce three steers and the farmer will pay the rancher not to produce the fourth and fifth steers. As we have seen, this will result in a Pareto-optimal state. But what if some external authority now steps in and imposes some form of wealth transfer from rancher to farmer? For many readers this would seem only fair. After all, it is the rancher who is the cause of the negative externality.[7] Assuming for the sake of argument that some form of *ex post* transfer is desirable, we now ask whether the resulting redistribution will be Pareto-optimal? That is, can the efficiency rationale be preserved? This is doubtful. Remember that (1) it is assumed that utilities are not intersubjectively comparable, and (2) no future trading possibilities are contemplated for the denizens of w_0. Given these assumptions, the notion of Pareto-optimality loses applicability. And if this is so, then we would seem to lack even a minimal sort of justification for an *ex post* scheme of wealth transfer. Perhaps this pessimism goes too far. And this because the prospect of an *ex post* adjustment would surely, at least in the real world, affect the initial bargaining stance of ranchers and farmers. (That is, a sense of Pareto-optimality could be restored by taking into account the effect possible *ex post* transfers will have on the initial bargaining stance.) But w_0 by construction does not allow for this sort of strategic consideration. And even if it did, there would still be a question of the (transaction) costs of such bargaining. So it is pretty clear that w_0 will have to be replaced with something more realistic. That's no surprise. What we do wonder, though, is whether the desirability of allocative efficiency, the schedule that calls for the production of three steers, will be retained in more realistic universes. Is this desirability a function of the particular idealizations we happen to be using, or is it independent of those idealizations and hence unproblematically applicable to the real world? In other words, will the recommendation that three steers be produced remain stable with respect to changes in idealization?

5. TAKING TRANSACTION COSTS INTO ACCOUNT

Rather than enter the realm of strategic bargaining, let us (following Coase 1960) increase the realism of w_0 by considering the effect transaction costs have on production schedules. This small increase in realism will be sufficient for my purposes. So by way of approaching the complexity of the real world, assume the existence of an idealized world where there are non-zero transaction costs but that is otherwise identical to w_0. Let this world be denoted by w_t.

For simplicity and ease of exposition, I shall assume a fixed transaction cost of $2 that is shared equally. Focus now on the case of the third steer. If there were no transaction costs, the marginal profit of the third steer to the rancher would be $30, and the marginal damage to the farmer would be $29. But given the assumed expenses of negotiation, the marginal profit to the rancher is $29, and the marginal damage to the farmer is $30. Consider now what happens in case the rancher is assigned production rights. If there were no negotiation, then he or she would make $30 if the third steer were produced. If there is negotiation, then the rancher would lose at least $1 if he or she were not to produce the third steer. This is because the lost marginal profit and transaction cost add up to $31, whereas the maximum rational offer that the farmer can make is $30. Therefore, in w_t, the third steer will be produced if production rights are assigned to the rancher. Consider next what happens if production rights are assigned to the farmer. In the case of no negotiation, the farmer saves $29 if the third steer is not produced. If there is negotiation and the third steer produced, then the farmer must lose money because the maximum offer the rancher can make of $29 is less than the sum of the marginal damages ($29) and the transaction cost ($1). Therefore, in w_t, the third steer will not be produced if production rights are assigned to the farmer.

So in w_t it matters how property rights are assigned. In particular, the production schedule in w_t will equal that of w_o *only if* the rancher has production rights *or* transaction costs are less than $1. There will be a lost surplus and market failure if farmers have production rights and transaction costs are greater than $1. Therefore, because we have been assuming that transaction costs are not less than $1, consensus will be in conflict with allocative efficiency if farmers have production rights. Giving farmers production rights means that allocative efficiency can be obtained only if the normal bargaining process between rancher and farmer is preempted in favor of some state or court imposed (approximation) of allocative efficiency. So in w_t there is a problem about what to do that does not exist in w_o. Utilitarian and deontic norms are now in conflict. Using w_o as our guide, our problems seemed to be (1) to determine who should have rights of production, and (2) to determine whether to impose some compensation scheme to effect *ex post* a transfer from the rights holders to non-rights holders. But, because of the emerging conflict between efficiency and consensus, w_t casts doubt on whether allocative efficiency should be kept as a goal for w_t, or more to the point, for the actual world. It depends–or so our discussion of w_o and w_t suggests–on how highly we value personal autonomy and consensus.

6. ROBUSTNESS

Our simple example illustrates a common phenomenon in economic analysis, *viz.*, a failure of robustness: predictive and normative consequences do not remain stable given increasingly more realistic assumptions.[8] This lack of robustness leads to a *dilemma* for those who wish to draw normative lessons from such idealized worlds. In order to see this, imagine that w_t has been made more realistic, say by imposing some game theoretic analysis of bargaining, or by including, in some way or other, third party effects. Call this new more realistic set of idealizations w_g. Now, by and large, bargained for production schedules will differ in w_g from those bargained for in w_t. But such differences lead to the following dilemma. On the one hand, if normative lessons are not modified to reflect more realistic worlds such as w_g, then such lessons are clearly arbitrary because they are tied to a particular set of idealizations which is just one among many available. On the other hand, if normative lessons are modified to reflect w_g, or even more realistic worlds, the lessons again appear arbitrary because they depend on which analyses are accidentally available to us (i.e., which possible worlds are contemplated).[9] So the problem for supporters of welfare economics is to find a way to circumvent the lack of robustness. Two possibilities suggest themselves. First, to show that robustness (or convergence) does appear once a sufficiently realistic plateau of idealization is reached. Second, to make some sort of end-run around the problems caused by a lack of robustness.

I shall not attempt to appraise prospects for the first possibility here other than to note that prospects for success seem dim. In science and engineering, controlled experimentation plays a crucial role in the development of theories that are likely to be true or successful. Generally speaking, there is an attempt to enumerate, and rank in terms of importance, the causal influences at work in situations of interest. Exactly how this is accomplished is an area of ongoing research in the philosophy of science.[10] But insofar as experimentation, well-controlled or otherwise, is impossible in normative considerations, the methods employed in science and engineering will not be applicable. Another technique used in science and engineering is to formally prove that stability exists for increasingly realistic analyses. A standard example is Newton's proof that treating a spherical object as if its mass were concentrated at its center is equivalent, for a wide class of problems, to treating the body as extended. In terms of the distinctions made above in section 2, what Newton did was to convert the idealization of treating an extended body as if it were a point mass into a bounded approximation. But such successes, even when available, are dependent on their own associated idealizations. Very important in the history of science are formal proofs showing that theories cannot be improved regardless of the realism of the idealizations used in their application. The classic example is that of kinetic theory and specific heats. As models became more elaborate, specific heat predictions necessarily diverged from experimental values because of the increase in degrees of freedom of motion. This case forms one of Kelvin's "two clouds" over turn-of-the-century science.[11] But even if possible, such proofs serve only to refute not to confirm theories.

To summarize, the prospects for showing of any idealized normative theory that it will be stable with respect to changes in idealization toward greater realism seem dim. It would seem to behoove us therefore to attempt some sort of end-run around the problem of stability as it exists in normative argumentation.

I shall now consider one popular way of making the required sort of end-run. Return to w_t. As explained, the production schedule in w_t will equal that of w_0 only if the rancher has production rights (i.e., return, for the moment, to our original formulation of the problem wherein one of the parties gets production rights and there are no *ex post* payments). Therefore, an institutionalized practice of affording production rights to ranchers will insure a resulting harmony between consensus and efficiency. That is, once production rights are assigned to ranchers, agents will freely bargain (in w_t) to efficient production schedules. Therefore–or so the argument goes–production rights ought to be granted to ranchers and not farmers. Otherwise the agreement between utilitarian and deontic goals will be lost. At first hearing this sounds good. But even ignoring problems due to the probable lack of robustness, there is no rationale (restricted to the terms of w_t) for why the resultant harmony between deontic and utilitarian goals should be allowed to so dominate public policy. After all, once ranchers are given production rights, what's up for negotiation is narrowly restricted. (As illustrated in Figure 2, negotiation is restricted to bargaining moves from what would have occurred if the fifth steer were produced.) A clever way around this problem, along Rawlsian lines, has been suggested by Richard Posner. (See Posner 1980; for critical discussion see Coleman 1980a, 1980b.) Assume that the transaction expenses of *ex post* payment schemes are prohibitive. That is, assume that the costs of administering a compensation system exceed marginal profits. As an initial idealization there's something to be said for such an assumption.[12] Let w_t as modified by this assumption be denoted w_t'. Now transform the problem from one of determining what rational parties should do in w_t' to the problem of determining what parties not in w_t' but about to enter will decide the institutional practices of w_t' should be. In effect, the strategy is to further escalate the level of counterfactuality. Posner's claim is that parties acting in an initial state of ignorance or uncertainty as to whether they will be farmers or ranchers will (if they are rational) decide on an institutionalized setting that awards production rights to ranchers.[13] If this is correct, the harmony between consensus and efficiency is once again restored. But in a way that justifies a general institutional choice as opposed to a way that justifies only a narrow range of trading options. And because of this, there is an unequivocal normative basis for awarding production rights.

This is all very slick but upon reflection seems to do no more than get us out of the pan of a suspicious normative basis into the fire of an obscure counterfactual original state of ignorance. Even assuming the correctness of the assumption of prohibitive administrative costs for *ex post* compensation, there is the problem of justifying the truth of the counterfactual that rational parties would opt to make the institutional choice to award production rights to ranchers. If the requirement that parties be rational is read in a weak everyday sense, then the counterfactual is indeterminate as to truth value. If, on the other hand, the rationality requirement is read so as to entail the truth of the counterfactual, then rationality itself is being understood in an implicitly normative way. But if this is so, then Posner's argument serves merely to effect a change of idealized venue. This change of idealized setting, however, serves only to return us once again to our original problem, namely, to demonstrate stability of normative output. For even if our idealized ranchers and farmers were to opt for giving productive rights in w_t' to those who turn out to be ranchers, this says nothing about what would be decided by agents about to enter more realistic worlds under conditions of uncertainty. So once again ignorance about the robustness of our idealized models ruins the cogency of the normative argumentation just as such ignorance ruins the cogency of argumentation used in science and engineering.

7. CONCLUSION

To summarize, the use of idealizations in science and engineering causes problems concerning the truth of theories and the projectibility of combinations of theory and idealization. Explaining the success of science and engineering in the face of such problems is a daunting task that philosophers of science have yet to complete. Important components of this explanation consist of (1) controlled experimentation used as a means of determining causal influences, and (2) the use of robustness arguments demonstrating the stability of outcomes given increasingly more realistic idealizations. But however difficult the problems may be in science and engineering, both for philosopher and working practitioner, the situation in normative argumentation appears worse. Controlled experimentation is conceptually impossible and stability arguments seem nowhere to be found. Proposed end-runs in terms of what rational agents would decide in suitable states of ignorance do nothing to answer concerns about stability. One strategy not discussed in this paper is that of naturalizing ethics. That is, to bring it within the scientific fold, perhaps under the auspices of evolutionary biology. While no cure-all, such an approach would make the consequences–conceived abstractly–of using idealizations no worse than they are in science and engineering.

Ohio State University

NOTES

[1] I have in mind here the use of dimensional analysis. See Laymon 1991 for the interplay between dimensional analysis and the use of idealizations.

[2] For more on the distinction between idealizations and approximations see Laymon 1987 and 1990. For a basic overview of the problems caused by idealizations for science see Laymon 1985 and 1989b. For a discussion of the problems caused by idealizations for engineering see Laymon 1989a and 1991.

[3] I hasten to emphasize that because I am a philosopher of science and not a value theorist my knowledge of the normative literature is not extensive. Some indirect support for my appraisal is to be found in Skyrms 1993, which contains a very interesting discussion of the effects the consideration of dynamical possibilities has on some standard uses made of equilibrium in game theory. As Skyrms shows, one simply cannot take for granted that static states of equilibrium will be dynamically possible. In certain respects, Skyrms's discussion is a reprise of the nineteenth and early twentieth century development of stability and control theory. Because elementary game theory plays such an important role in contemporary discussions of social and political philosophy, Skyrms's results show the vulnerability of many of the claimed philosophical results.

[4] Balzer *et al.*, 1987, p. 324. I have attributed the quotation exclusively to Moulines because he is the author of the chapter in question.

[5] Basic readings on the economic analysis of law include Ackerman 1975, Coase 1960, Coleman 1988, Kennedy and Michelman 1980, Kronman and Posner 1979, Polinsky 1983, and Posner 1973.

[6] The example originates in Coase 1960.

[7] For a discussion of the relevance of the causal aspects of this example see Coleman 1980a, pp. 79-81.

[8] For more on robustness see Gibbard and Varian 1978, and Laymon 1993.

[9] There is the additional difficulty that the set of considered models or possible worlds may not be linearly ordered, in which case there will not be a most realistic description.

[10] For a very suggestive discussion of the experimental study of causation see Cartwright 1989.

[11] See Thomson 1901. See Laymon 1988 for a discussion of Kelvin's other cloud, the Michelson-Morley experiment, and of Lorentz's proof that increasingly more realistic analyses could not be made compatible with the null result.

[12] We have in this paper simply assumed the existence of a profit and damages schedule for rancher and farmer. But how such a schedule is to be

reliably obtained in the real world, i.e., in a world where strategic bargaining is a reality and where there are inducements to exaggerate profits and damages, is a difficult problem. Posner's assumption that the costs of administering a compensation system exceed marginal profits gains credence insofar as one believes such problems to be insurmountable. For an introduction to "demand-revealing" processes see Tideman 1977.

13 Because of space limitations, I've had to simplify Posner's position here.

REFERENCES

Ackerman, Bruce (1975). *The Economic Foundations of Property Law*. Boston: Little, Brown.

Ammann, O., von Karman, T., and Woodruff, G. (1941). "The Failure of the Tacoma Narrows Bridge: A Report to the Honorable John M. Carmody, Administrator, Federal Works Agency." Unpublished. Reprinted in *Bulletin of the Agricultural and Mechanical College of Texas* 15 (1944). There is no overall pagination in this issue of the Bulletin; all page references are to those of the original reports.

Andrew, C., Gregory, L., and McMinn, R. (1938). "Report to the Board of Consulting Engineers, The Tacoma Narrows Bridge, to Mr. Lacey V. Murrow, Director of Highways." Unpublished. Reprinted in *Bulletin of the Agricultural and Mechanical College of Texas* 15 (1944).

Balzer, C., Moulines, U., and Sneed, J. (1987). *An Architectonic for Science*. Dordrecht: Reidel.

Coase, Ronald (1960). "The Problem of Social Cost." *Journal of Law and Economics* 3: 1-33.

Coleman, Jules (1980a). "Efficiency, Auction and Exchange," *California Law Review* 68: 221-252. Reprinted in Coleman 1988, pp. 67-94.

Coleman, Jules (1980b). "Efficiency, Utility and Wealth Maximization," *Hofstra Law Review* 8: 487-507. Reprinted in Coleman 1988, pp. 95-132.

Coleman, Jules (1988). *Markets, Morals and the Law*. New York: Cambridge University Press.

Condron, T. (1938). "Excerpts from Report of Supervisory Engineer, to Washington Toll Bridge Authority of the State of Washington." Unpublished. Reprinted in *Bulletin of the Agricultural and Mechanical College of Texas* 15 (1944).

Kennedy, Duncan, and Michelman, Frank (1980). "Are Property and Contract Efficient?" *Hofstra Law Review* 8: 711-729.

Kronman, Anthony T., and Posner, Richard (1979). *The Economics of Contract Law*. Boston: Little, Brown.

Gibbard, Allan, and Varian, Hal R. (1978). "Economic Models." *The Journal of Philosophy* 75: 664-677.

Laymon, R. (1985). "Idealizations and the Testing of Theories by Experimentation." Pp. 147-173 in *Experiment and Observation in Modern Science*, edited by P. Achinstein and O. Hannaway. Boston: MIT Press and Bradford Books.

Laymon, R. (1987). "Using Scott Domains to Explicate the Notions of Approximate and Idealized Data." *Philosophy of Science* 54: 194-221.

Laymon, R. (1988). "The Michelson-Morley Experiment and the Appraisal of Theories." Pp. 245-266 in *Scrutinizing Science: Empirical Studies of Scientific Change*, edited by L. Laudan, R. Laudan, and A. Donovan. Dordrecht: Kluwer.

Laymon, R. (1989*a*). "The Application of Idealized Scientific Theories to Engineering." *Synthese* 81: 353-371.

Laymon, R. (1989*b*). "Cartwright and the Lying Laws of Physics." *Journal of Philosophy* 86: 353-372.

Laymon, R. (1990). "Computer Simulations, Idealization and Approximation." Pp. 519-534 in *PSA 1990* (Proceedings, Philosophy of Science Association), v. 2, edited by A. Fine, M. Forbes and L. Wessels. East Lansing: Philosophy of Science Association.

Laymon, R. (1991). "Idealizations and the Reliability of Dimensional Analysis." Pp. 146-180 in *Critical Perspectives on Nonacademic Science and Engineering*, edited by Paul Durbin. Bethlehem: Lehigh University Press.

Laymon, R. (1993). "The Confirmational and Computational Differences between the Social and the Physical Sciences." *Philosophia* 22: 241-273.

Lin, C. C., and Segel, L. A. (1974). *Mathematics Applied to Deterministic Problems in the Natural Sciences*. New York: Macmillan.

Pippard, A., Tuplin, W., and McEwen, E. (1953). "Your Reviewer," "Letters to the Editor." *The Engineer* 196: 369-370, 465-466, 561. Reprinted pp. 187-196 in *Contributions to a Philosophy of Technology*, edited by F. Rapp. Dordrecht: Reidel, 1974.

Polinsky, A. Mitchell (1983). *An Introduction to Law and Economics*. Boston: Little, Brown.

Posner, Richard (1973). *The Economic Analysis of Law*. Boston: Little, Brown.

Posner, Richard (1980). "The Ethical and Political Basis of the Efficiency Norm in Common Law Adjudication." *Hofstra Law Review* 8: 487-507.

Skyrms, Brian (1993). "Chaos and the Explanatory Significance of Equilibrium: Strange Attractors in Evolutionary Game Dynamics." Pp. 374-396 in *PSA 1993* (Proceedings, Philosophy of Science Association), v. 2, edited by D. Hull, M. Forbes, and K. Okruhlik. East Lansing: Philosophy of Science Association.

Tideman, T. Nicolaus. "Introduction." *Public Choice* 29:2: 1-15.
Thomson, W. (Lord Kelvin) (1901). *Baltimore Lectures on Molecular Dynamics and the Wave Theory of Light*. London: C. J. Clay.

LARRY A. HICKMAN

TECHNIQUES OF DISCOVERY: BROAD AND NARROW CHARACTERIZATIONS OF TECHNOLOGY

Historical studies of technology have tended to exhibit two focal points. Their primary concern has usually been what has happened or happens in the domain of hardware such as tools, machines, structures, and even items of personal use. Second (and secondarily), they have concerned themselves with what were or are the "attitudes" that accompanied the invention, development, or use of such hardware. Such secondary considerations have included the cultural matrices that made such inventions and developments possible and the cultural consequences of their employment. One of the ways that historians have expressed this difference of emphasis has been to identify the stock of hardware available for use during a given historical period as its "civilization," or "tools," and its ideas and institutions as its "culture."

Corresponding to their different ways of weighting these two factors, histories of technology may be said to represent a spectrum of assumptions about how technology is to be characterized. On one end of the spectrum are narrow treatments, such as some of those found in the journal, *Technology and Culture*, for which technology is more or less identified with hardware. Typical issues of that journal include essays about threshers, steam engines, and even the steam powered privy pump. The journal's title indicates one of its working assumptions: on the one hand there is "technology" (one is tempted to read this as "civilization"), and on the other hand there is "culture."

Somewhat nearer the middle of the historian's spectrum are studies that treat technology in ways that attempt to bring various cultural matters under the tent. In *Technics and Civilization* (1934), for example, Lewis Mumford provided his readers with extended accounts of the invention and development of the hardware of production, and he even took into account the introduction of glass mirrors and other personal or "end-user" artifacts. But in all this his readers are never more than a few pages away from a detailed

Joseph C. Pitt (ed.), New Directions in the Philosophy of Technology, 207-218.

discussion of the attitudes that conditioned such innovations and the ways in which their introduction and use modified the cultures into which they had been introduced. The same may be said of the work of Lynn White, Jr., as for example, in his well known discourse on the relative merits of the button and the sonnet (White 1963). For both of these historians, what was cultural was to some extent treated as a part of broader technological phenomena.

In his classic and still impressive work, *Mechanization Takes Command* (1948), Siegfried Giedion, too, sought to narrow the gap between the hardware and the cultural components of technology. His unique insight was that, perhaps, even to a greater degree than with the products of celebrated inventors such as Edison and Marconi, it has been anonymous inventions and the cultural climates that supported them that have broadly and profoundly influenced the course of human life. Giedion was fascinated, for example, by the intellectual climate that gave rise to "the harnessing of vapors, steam, and gases to unwonted ends" (Giedion 1948, p. 185), including hot air balloon railways up mountain slopes, machines for producing soda water and bread, and the Bessemer method for producing steel. Nevertheless, he continued to give grudging respect to the implicit division of matters into "technology" on the one side and "attitudes toward technology" on the other. For all its obvious merits, there remained in Giedion's work a perceptible gap between tangible hardware and intangible culture.

More recently, historian of literature Cecilia Tichi, in *Changing Gears: Technology, Literature and Culture in Modernist America*, took deliberate and decisive steps to narrow this gap. A technological revolution, she announced, "is a revolution not only of science and technology, but of language, of fiction, and ultimately of poetry" (Tichi 1987, p. 16). The idea that she articulated more forcefully than Mumford or Giedion ever had is that careful analyses of technological innovations make it difficult to justify any hard and fast separation between components that are tangible and ones that are intangible. Her implicit argument was that the very distinction is delineated in terms of a particular set of questions under discussion, and is itself artifactual.

Some philosophers have moved even further in the direction of supplanting, or at least relativizing, the hoary distinction between the tangible and the intangible. Recent essays by Lester Embree and Carl Mitcham provide excellent examples. In his essay, "A

Perspective on Rationalistic Technology or How to Buy a Car," for example, Embree argues that "although the word 'technology' often merely signifies equipment . . . a modicum of reflection shows that things are equipment only in relation to how they are used. If 'equipment' can then be used to signify equipment, 'technology' can signify the *using of equipment*" (Casey and Embree 1989, p. 146).

Among the things Embree counts as equipment are such putatively non-artifactual objects as the sun and the stars for navigators and astrologers, and even the ideas put into play by a teacher such as Socrates whose classroom was an Athenian street and whose only tools were a stick, a plot of sand, the figures scratched into it, and, of course, his ideas. With the possible exception of Tichi, Embree thus treats technology much more broadly than any of the writers I have so far mentioned. In his view, technology includes a noetic or conceptual component that is at least as important as its tangible hardware component.

Carl Mitcham takes much the same tack in his essay, "On Going to Church and Technology" (Casey and Embree 1989, pp. 221-236). Since going to church is an intentional matter that involves special clothing, attitudes, and preparations, and since specific outcomes are expected and worked toward, Mitcham sees no difficulty in broadening the notion of technology to include such things as polishing one's shoes, taking care not to step on the oil spots in the church parking lot (because the soles of one's shoes will be on display as one kneels at the altar), and even mental preparations for paying attention to the sermon. Each of these elements of going to church is clearly artifactual, he argues, because each is under the direct control of someone who seeks to utilize it as a tool in the performance of some task in order to achieve some end which he or she considers desirable.

In this matter Mitcham and Embree are representatives of an approach to technology studies that may be called "phenomenological." In their view, careful attention to the phenomenology of everyday affairs is regarded as an indispensable element of any thorough study of technology. Furthermore, everyday affairs are so defined that they include what we variously call the noetic, conceptual, ideational, or "mental." Their implicit claim is that there is no *prima facie* reason for distinguishing tangible from intangible, or outer from inner, when the goal is an understanding of human activity. In their view there may, however, be good grounds *within* a particular sequence of inquiry for making such a distinction. Implicit in their approach is the view that the distinction between tangible and intangible is itself functional and artifactual–a tool for use in a given bit of inquiry. This is a view that I find quite salutary for reasons that I will indicate a few paragraphs hence.

To some, this phenomenological approach may appear to constitute a kind of overkill. It may seem that Mitcham and Embree have defined technology so broadly that it includes virtually everything. Embree himself seems a bit worried about this possible consequence, and issues the following qualification: "One might think that this line of thinking leads to everything being technological. But equipment use pertains to a *practical attitude* in which some aspect of the course of the world is at issue, which is not the case in attitudes in which things are simply enjoyed or in which the goal is knowledge. . . . Nevertheless, it seems wiser to begin with the technological and search for the nontechnological than vice versa" (Casey and Embree 1989, p. 147).

In other words, Embree wants to emphasize the noetic as a part of the technological because most of our practical activities involve manipulation of our environment in ways that cannot be accomplished in the absence of some measure of noetic contribution. At the same time, however, he pointedly excludes from the technological both what is unreflectively enjoyed *and* what is theoretical.

I believe that the first half of Embree's position–that technology should be defined broadly enough to include the noetic but not what is immediate or unreflective–not only reflects actual human technological activity; it also has significant beneficial consequences. An obvious reason why we should "begin with the technological and search for the nontechnological," as he puts it, is because there are good grounds to identify the cognitive with the technological. This is because wherever the noncognitive phase of experience is brought into focus and reconstructed, it is the cognitive phase of experience that does the work. If the cognitive is that portion of experience that does work and produces consequences, and if the noncognitive is that portion of experience that is simply enjoyed or suffered immediately and unreflectively until it becomes raw material for cognition, then the first half of Embree's claim is secure.

The same considerations, however, render the second half of Embree's position problematic. This is because theoretical knowing, as well as practical knowing, is an integral part of cognitive activity. It is therefore difficult to share Embree's worries that we go too far if we include as technological what is noetic in the sense of the theoretical. Once cognition is understood as productive, or technological, then the practical and the theoretical fall into place as forms of technological activity: both types of knowing cooperate in the invention, development, and use of tools and other artifacts in order to achieve certain desirable ends or goals.

To take what may be the most extreme case imaginable, I would argue, for example, that a logician seated in a room alone and without pencil, paper, or computer, but engaged in the solution of some favorite but irritating problem, is involved in technological activity. In this case she is using certain tools, probably including variables, numbers, and transformation rules. She is performing certain operations with those tools just as surely as a carpenter performs certain operations with hammers, saws, and squares in the building of a garage and just as surely as a mechanic performs certain operations with wrenches, lubricants, and fasteners in the building of an automobile engine. In each case there is also a desired outcome: a proof, a garage, an engine.

One of the unfortunate consequences of yielding to Embree's concern to exclude theoretical knowing from the domain of technological activity is a divorce of what is "scientific" from what is "technological." This view, that science is fundamentally different from technology, has been a longstanding and popular view, and one that has been held even by more analytically inclined philosophers such as Mario Bunge. Technology, on this view, is a practical matter that involves the application of scientific theory. On this view, the philosophy of technology becomes "the study of rules–grounded rules of applied science." As Mitcham and Mackey characterize Bunge's view, he is "concerned with analyzing the logic of technological rules [and] determining exactly how they are grounded in scientific laws" (Mitcham and Mackey 1972, p. 4; Bunge 1967).

It seems to me that this view ignores important data. In fact, the activities of almost all scientists, including not only those who work in the "wet" labs of the chemists, but also those whose work may include years of calculations which are put to the test during a few days at a particle accelerator, involve the use of instruments that are "technological" even in the thinnest and least controversial of senses. The beakers and spectrometers of the chemist, the telescopes of the astronomer, and the cyclotrons of the physicist are all patently technological artifacts.

This is in fact what separates contemporary science from what passed for science among the Athenian Greeks. The "science" of Aristotle was empirical and only minimally instrumental and therefore minimally intrusive into natural phenomena; it was primarily a matter of observing, describing, and cataloging. Contemporary science, on the other hand, is instrumental and intrusive at its core; it is experimental in the sense that it requires the use of instrumentation to change situations, not just to record them. Another way of putting this is that contemporary science is technological (in the sense of being dependent on the use of hardware artifacts) in a way that classical science never was.

At the far reaches of theoretical science, of course, as in the case of the logician sitting alone in her room, matters are more controversial. A commitment to the view that science operates without technological admixture in such cases often includes a commitment to a type of ontological realism that was made popular by Plato and later reinforced by Gottlob Frege, namely that the objects of logic are supernatural or natural in origin and therefore just "found" or discovered. On this view they are not artifactual and therefore not constructed. This view in its turn requires an acceptance of another view, namely that the origins of logical objects are not amenable to discovery since their existence is somehow eternal.

My own reaction to this view is that it is a very heavy price to pay for an attempt to purify what is "theoretical" or "scientific" of what is "technological," since it merely stipulates a view of logical and mathematical objects that locates them in a realm that is by definition beyond the boundaries of inquiry into their provenance. In short, this view blocks the road to further inquiry into the origins of logical and mathematical tools. The price of ignoring the evolution of logical and mathematical tools is a *de facto* Platonism, which in turn has its roots in Pythagorean and other varieties of mysticism.

A far more productive way of thinking about these matters, I would argue, is to think of an abstract science such as formal logic or mathematics as a type of technological enterprise. It is the type of technology that operates primarily with tools and other artifacts that are primarily or for the most part noetic, that constructs and reconstructs the relations among such noetic objects, and that is thereby capable of providing platforms for the construction of further meanings of those and other objects, including the existential ones against which they may ultimately be checked.

This view has several advantages. First, it validates a relatively new but potentially fecund area of inquiry, namely, the genetic analysis of these noetic tools. Just as there is a vestige in the modern plow of the bent stick, there is a vestige in the square root of minus one of the marks made on the wall of an ancient shepherd's fold in order to compare the number of outgoing sheep in the morning to the number of incoming sheep in the evening (to take just one, somewhat speculative example). And it is hardly surprising that organisms with ten fingers, counting thumbs, would tend to develop number systems of base ten.

This genetic approach argues that the noetic tools of science are constructed, but not that they are constructed out of nothing. When they are sophisticated and complex they are constructed out of previous tools and other artifacts. They are relations of relations. And the most primitive of such tools are constructed out of the rawest of raw empirical materials: felt needs and desires, flashes of insight, and accidents.

Another advantage to the view I am advancing is that it leads us to look for continuities between the adjustive activities of human beings and the adjustive activities of other natural organisms. If science is a kind of technology, then technology, as the sum of ways that human beings accommodate themselves to their environments and alter those environments to their needs, serves as a kind of linkage or bridge to similar activities undertaken by higher primates, and even by "lower" animals.

The operative concept in this context is *habit*. Viewed as a part of a larger picture, an artifact is a kind of habit. It is something that has a certain generality of application; it is something that has been tried out and found to be capable of serving a range of purposes. Seen in this way, as habits of a sort, hammers and saws become continuous with the other habits developed over millennia by higher order primates, for example, in their attempts to adjust to changing environmental conditions. Seen in this way, to say that human beings are uniquely technological animals is not to place them outside and above nature, but within nature and as a part of it. Their activities differ from those of their nonhuman relatives and forebears not in kind, but in level of complexity. Habits are found throughout nature, but only human beings have reached the level of complexity that allows a high level of self-control with respect to their formation, development, retention, and modification. It is for this reason–their ability to engage in the self-controlled manipulation of habits–that human beings are able to reach very high levels of efficiency.

There is a second, and I think unintended, consequence of Embree's divorce of theoretical knowing from knowing that is technological. If we do not count theoretical knowing as artifactual and therefore as technological, then we continue to be entrapped in a kind of epistemological dualism that has been around since Plato and Aristotle.

This is the view that knowledge is the reception within consciousness of external forms, or the attenuated replication within consciousness of preexisting, fixed "essences." In other words, this is the view that there is an external, self-sufficient, and fully formed reality that exists anterior to our knowledge of it. Some have called this variety of realism the "mirror" theory of reality.

The mirror theory suffers from several disadvantages. First, it renders knowing passive rather than active, since knowing is more a matter of taking something in than of the use of the raw materials of experience in the construction or reconstruction of some outcome. Second, it is conservative, rather than progressive, since it leads to the view that the world that corresponds to our knowledge is already just as we will determine it to be, that is, that knowing it cannot alter it except in the trivial sense that it becomes known, or mirrored. Knowing is thus rendered less effective for improving matters than would otherwise be the case, and the experimental is depreciated.

Improvement, or amelioration, since it is not a part of the theoretical, is on this view relegated to the domain of the practical. (On this model, for example, a scientist whose work is utilized in a way that ultimately leads to the production of high-level nuclear waste is justified in saying that she is just doing her theoretical work, i.e., finding out about the "structure" of the universe, and that the problem of waste is just a practical matter and therefore none of her concern.) Third, this view often depends on the old and unfortunate notion that science deals with certainty. It is only with considerable sleight-of-hand that science, on this view, is able to come to terms with the newer view that probabilism and fallibilism are the hallmarks of scientific technology.

Corresponding to these three disadvantages of the older dualism are three advantages of accepting theoretical knowing as a technological activity and theoretical knowledge as a type of technological artifact.

First, the work involved in the getting of scientific knowledge – even the knowledge of those scientific laws that philosophers such as Bunge think ground technology – becomes itself a matter of active and continuing reconstruction. Take, for example, current theories of brewing beer. To be sure, there is an important sense in which the laws of chemistry play a part in the work of the brewmaster. To be sure, chemists have determined that elements combine in certain ways to form certain compounds, and this knowledge can aid the brewmaster in the improvement of his or her product. And so it may appear to some that the case just involves the practical application of theoretical knowledge: an application by the brewmaster of certain brute facts about the proclivities of chemical substances to combine in certain ways. On this view, the brute facts are said to have been discovered by the chemist.

But it seems unlikely either that the information thus utilized by the brewmaster was always recognized as brute fact or that such recognition was just a matter of observation. In order to become a part of the body of knowledge we call chemistry, experimentation was required, and this involved the construction of the tools we call hypotheses and the development of other instrumentation for testing. What appears as a brute fact to one generation of chemists may have been a matter of considerable difficulty to an earlier generation. New constructions had to be undertaken, and new artifacts had to be produced.

To use one of Embree's examples, the sun experienced by our less sophisticated forebears is not the same sun as is experienced by contemporary navigators and astrophysicists. But even though it is not the same object, neither is it a totally different one. The sun of the contemporary astrophysicist is to the sun of his neolithic predecessor as modern plows are to bent or sharpened sticks. There is both continuity and difference.

To put matters succinctly, the "mirror theory" realist looks backward at work previously done and exhibits a certain arrogance. Because of his narrow definition of technology, he ignores the hard work that has gone into the construction of what he calls "the facts." He does not see that facts are themselves constructed or artifactual, and not just given. Facts are always facts-of-a-case. Even the selection of these facts from an indefinitely extended field of possible attention constitutes an act of reconstruction of that field. On the other hand, because of her more inclusive definition of technology, the constructivist looks forward at work to be done, and simultaneously exhibits a deeper respect for the work of her forebears. Her piety is the result of her recognition of the importance of genetic accounts of tool-development and her sensitivity to the pervasiveness of artifactuality.

Second, the constructive position, the view that all knowledge–whether theoretical or practical–is artifactual, has the advantage of providing secure and steady platforms for the improvement of situations that are not as we wish them to be. It is not that we "look for" solutions, but that we seek to construct them. Knowing is not so much a matter of "finding out" as it is a matter of "making sure." On this view, the kind of inquiry that leads to greater control of problematic social and political situations is also a type of technological undertaking, since it involves an active construction of desirable outcomes through the use of the tools and artifacts that are proper to that domain of knowledge-getting. Not only science itself, but the sociology and politics of science become important technological undertakings.

Third, if knowing is a technological artifact, then certainty is replaced with dependability. High levels of dependability are restricted to certain narrow domains that exhibit a finite number of cases or relations and in which sufficient work has been done to insure that further reconstruction of the tools and artifacts in use there will be minimal. The laws of mathematical addition and subtraction are dependable in this sense not because they correspond to "the furniture of the world," to use Bertrand Russell's infelicitous phrase, but because a great deal of work has been focused on a very narrow area of inquiry and because the connections between the artifacts in that area and the domain of existential affairs have been intentionally limited. As for the remaining domains of inquiry, which constitute the vast majority of the locations where scientific-technological work is done, reconstruction continues to be done on the assumption that further improvements can be made. "Fallibilism," "probability," "dependability," and "warranted assertibility" replace "certainty" as key operational terms.

What I have attempted to do in the preceding material is to suggest some of the reasons why I think that broad-based characterizations of technology have considerable advantages over those that are more narrowly based. But I still need to respond to those who fear that if my view is accepted then "technology will just have to do with everything."

I have characterized technology as the production and use of tools of all sorts, both tangible and intangible or noetic, to perform specific tasks, in order to achieve certain desirable ends or goals. I have argued that what is commonly called "theoretical science" is no less a type of this activity than is hardware technology. Theoretical knowing is no less a case of technological activity than is knowing that is involved with concrete, practical outcomes. Further, because the theoretical is also artifactual, science is a type of technology. Is there anything, then, that does not count as technological?

I believe that the first half of Embree's suggestion–that most of what we do on a daily basis is not technological, although it most often exhibits certain technological overtones and is possible only because of technological work already done, either by ourselves or by someone else–goes a long way toward answering this question. Most of the things we experience and do on a daily basis involve rote activities, unreflective habitual responses, and immediate and uncritical aesthetic enjoyments. Most of what we experience and do does not involve knowing in the sense in which I have developed and utilized the term. I have, however, already presented my reasons for rejecting the second half of his suggestion, namely that the theoretical is not technological.

So whereas the narrow characterizations of technology utilized by many historians often tend to draw a line between hardware and everything else artifactual, which is then commonly called culture, and whereas phenomenological accounts such as Embree's often tend to draw a line between the practical and the theoretical, I want to draw a line between what is involved in and a conscious result of intelligent reconstructive activity, on one side, and what is merely passive, rote, and does not involve inquiry, on the other. It seems to me that by dividing things up as I have we achieve a kind of continuity between science and technology that increases our power to effect meaningful adaptive change, that we are able to develop a wider appreciation for the ways that human beings function in and as a part of nature, and we are able to see the relevance of and make more sense out of genetic or historical studies.

Southern Illinois University at Carbondale

REFERENCES

Bunge, Mario. "Toward a Philosophy of Technology," in *Philosophy and Technology*, ed. Carl Mitcham and Robert Mackey. New York: Free Press, 1972, pp. 62-76. This essay is adapted from chapter 11, "Action," in Mario Bunge, *Scientific Research II: The Search for Truth*, vol. 3, part 2 of *Studies in the Foundations, Methodology, and Philosophy of Science*. Berlin, Heidelberg, New York: Springer-Verlag, 1967.

Embree, Lester. "A Perspective on the Rationality of Scientific Technology or How to Buy a Car," in *Lifeworld and Technology*, ed. Timothy Casey and Lester Embree. Washington, D.C.: The Center for Advanced Research in Phenomenology and University Press of America, 1989, pp. 145-63.

Giedion, Siegfried. *Mechanization Takes Command.* New York: W. W. Norton, 1948.

Mitcham, Carl. "On Going to Church and Technology," in *Lifeworld and Technology*, pp. 221-36.

Mitcham, Carl, and Robert Mackey, eds. *Philosophy and Technology.* New York: Free Press, 1972.

Mumford, Lewis. *Technics and Civilization.* New York: Harcourt, Brace & World, 1934.

Tichi, Cecilia. *Shifting Gears.* Chapel Hill: University of North Carolina Press, 1987.

White, Lynn. "The Act of Invention," in *The Technological Order: Proceedings of the Encyclopaedia Britannica Conference*, ed. Carl Stover. Wayne State University Press, Detroit, 1963, pp. 102-116.

NAME INDEX

PHILOSOPHY AND TECHNOLOGY

Series Editor: Paul T. Durbin

OFFICIAL PUBLICATIONS OF THE SOCIETY FOR PHILOSOPHY AND TECHNOLOGY

1. *Philosophy and Technology*
   Edited by Paul T. Durbin and Friedrich Rapp. 1983 ISBN 90-277-1576-9
   (Published as Volume 80 in 'Boston Studies in the Philosophy of Science')
2. *Philosophy and Technology, II.* Information Technology and Computors in Theory and Practice.
   Edited by Carl Mitcham and Alois Huning. 1986 ISBN 90-277-1975-6
   (Published as Volume 90 in 'Boston Studies in the Philosophy of Science')
3. *Technology and Responsibility*
   Edited by Paul T. Durbin. 1987 ISBN 90-277-2415-6; Pb 90-277-2416-4
4. *Technology and Contemporary Life*
   Edited by Paul T. Durbin. 1988 ISBN 90-277-2570-5; Pb 90-277-2571-3
5. *Technological Transformation.* Contextual and Conceptual Implications
   Edited by Edmund F. Byrne and Joseph C. Pitt. 1989 ISBN 90-277-2826-7
6. *Philosophy of Technology.* Practical, Historical and Other Dimensions
   Edited by Paul T. Durbin. 1989 ISBN 0-7923-0139-0
7. *Broad and Narrow Interpretations of Philosophy of Technology*
   Edited by Paul T. Durbin. 1990 ISBN 0-7923-0684-8
8. *Europe, America, and Technology: Philosophical Perspectives*
   Edited by Paul T. Durbin. 1991 ISBN 0-7923-1254-6
9. *Democracy in a Technological Society*
   Edited by Langdon Winner. 1992. ISBN 0-7923-1995-8
10. *Philosophy of Technology in Spanish Speaking Countries*
    Edited by Carl Mitcham. 1993 ISBN 0-7923-2567-2
11. *New Directions in the Philosophy of Technology*
    Edited by Joseph C. Pitt. 1995 ISBN 0-7923-3661-5

KLUWER ACADEMIC PUBLISHERS – DORDRECHT / BOSTON / LONDON

GPSR Compliance

The European Union's (EU) General Product Safety Regulation (GPSR) is a set of rules that requires consumer products to be safe and our obligations to ensure this.

If you have any concerns about our products, you can contact us on ProductSafety@springernature.com

In case Publisher is established outside the EU, the EU authorized representative is:

Springer Nature Customer Service Center GmbH
Europaplatz 3
69115 Heidelberg, Germany

Batch number: 10584442

Printed by Printforce, the Netherlands